Introducing
GLOBAL ISSUES

Introducing

GLOBAL ISSUES

edited by
Michael T. Snarr
D. Neil Snarr

LYNNE
RIENNER
PUBLISHERS

BOULDER
LONDON

Published in the United States of America in 1998 by
Lynne Rienner Publishers, Inc.
1800 30th Street, Boulder, Colorado 80301

and in the United Kingdom by
Lynne Rienner Publishers, Inc.
3 Henrietta Street, Covent Garden, London WC2E 8LU

Library of Congress Cataloging-in-Publication Data
Introducing global issues / edited by Michael T. Snarr and
    D. Neil Snarr.
    Includes bibliographical references and index.
    ISBN 1-55587-587-4 (hardcover : alk. paper). — ISBN 1-55587-595-5
(pbk. : alk. paper)
    1. World politics—1989–   2. International economic relations.
3. Social history—1970–   4. Ecology. I. Snarr, Michael T.
II. Snarr, Neil, 1933– .
D860.I62   1998
909.82—dc21   98-15207
                                                              CIP

British Cataloguing in Publication Data
A Cataloguing in Publication record for this book
is available from the British Library.

Printed and bound in the United States of America

5 4 3 2 1

For our wives, Melissa and Ruth,
whose patience and support deserve more credit
than can be acknowledged

Contents

Acknowledgments

We would like to express our appreciation to those who made this book possible. Jeffrey Lantis, John McLaughlin, Stephen Poe, Gerald Sazama, Amanda Dobbs, Michael Ebbert, Kimberly Hawk, Alison Johnson, Sekou Ade Mark, Christina Ralbovsky, and Kim Pavlina assisted us by reading and commenting on parts of the manuscript. Special thanks go to Margaret Degenhardt and Rena Hutton, who proofread several chapters. We are indebted to Connie Crecion for providing outstanding secretarial help and to Divya Thadani for technical assistance.

We would like to thank Lynne Rienner for her support of this project and Sally Glover who promptly answered the multitude of questions we asked. Of course, we also owe a great deal of thanks to the contributors to this book, who were patient with what must have seemed like a never-ending stream of requests.

We are also grateful to our institutions for support. Special thanks go to Wheeling Jesuit University, which provided release time through its Scholar-in-Residence program, and to the Social Science Department at Wheeling Jesuit, which offered unwavering support.

Most important, we would like to thank our families for bearing with us throughout this demanding process. Both our wives read significant portions of the manuscripts and gave valuable comments. The book would not have been possible without them.

Michael T. Snarr
D. Neil Snarr

Introduction

Michael T. Snarr

- Approximately 230,000 people are added to the world's population every day; that is the equivalent of 84 million people per year (Crossette 1996b).
- People in more than 130 countries have access to Cable News Network (CNN) (Barber 1996).
- In one region of Australia, a majority of the people over sixty-five years of age have skin cancer (Gore 1992).
- The number of individuals suffering from lack of food has declined over the past two decades (FAO 1996).
- Each year nearly 80,000 square miles of forest are depleted (with only a fraction of it being reforested). This is equal to the total territory of Maine, Massachusetts, and Virginia (Rourke 1997).
- Over the past two decades, the lives of 3 million children per year have been saved by immunization programs (UNDP 1996).
- More civilians have died this century as a result of war than in the four previous centuries combined.
- McDonald's and Kentucky Fried Chicken served more customers than any other restaurants in Japan in 1992 (Barber 1996).
- Tens of thousands of species are becoming extinct every year, and the rate is increasing.
- More than 1 billion people live in absolute poverty (UNDP 1996).
- Global military expenditures have decreased over the past decade (UNDP 1994).
- At the end of the century, 90 percent of the market for Coca-Cola will be outside the United States (Hauchler and Kennedy 1994).

- Nearly 20 million people are HIV-infected (UNDP 1996).
- Smallpox has been wiped out.

Each of the items above is related to a global issue discussed in this book. But what is a *global issue?* The term is used in the book to refer to two types of phenomena. First, there are those issues that cross political boundaries (country borders) and therefore affect individuals in more than one country. A clear example is air pollution produced by a factory in the United States and blown into Canada. Second, there are problems and issues that do not necessarily cross borders but affect a large number of individuals throughout the world. Ethnic rivalries and human rights violations, for example, may occur within a single country but have a far wider impact.

Our primary goal is to introduce several of the most pressing global issues and demonstrate how strongly they are interconnected. We also hope to motivate the reader to learn more about global issues and in turn to be a positive force for change.

■ IS THE WORLD SHRINKING?

There has been a great deal of discussion in recent years about globalization, which can be defined as "the intensification of economic, political, social, and cultural relations across borders" (Holm and Sørensen 1995: 1). Evidence of globalization is seen regularly in our daily lives. In the United States, grocery stores and shops at the local mall are stocked with items produced abroad. Likewise, Chicago Bulls, New York Yankees, and Dallas Cowboys hats and T-shirts are easily found outside of the United States. In many "foreign" countries, Madonna, Michael Jackson, Metallica, and other U.S. music groups dominate the airways; CNN and *Baywatch* are on televison screens; and Arnold Schwarzenegger is at the movies. Are we moving toward a single global culture? In the words of Benjamin Barber, we are being influenced by "the onrush of economic and ecological forces that demand integration and uniformity and that mesmerize the world with fast music, fast computers, and fast food—with MTV, Macintosh, and Mc-Donald's, pressing nations into one commercially homogenous global network: one McWorld tied together by technology, ecology, communication, and commerce" (Barber 1992: 53).

Technology is perhaps the most visible aspect of globalization and in many ways its driving force. Communications technology has revolutionized our information systems. "CNN . . . now reaches more than 140 countries" (Iyer 1993: 86); "computer, television, cable, satellite, laser, fiber-optic, and microchip technologies [are] combining to create a vast interactive communications and information network that can potentially give every person on earth access to every other person, and make every datum, every byte, available to every set of eyes" (Barber 1992: 58). Technology has also aided the increase in international trade and international capital flows and enhanced the spread of Western, primarily U.S., culture.

Of course the earth is not literally shrinking, but in light of the rate at which travel and communication speeds have increased, the world has in a sense become smaller. Thus, many scholars assert that we are living in a qualitatively different time, in which humans are interconnected more than ever before. "There is a distinction between the contemporary experience of change and that of earlier generations: never before has change come so rapidly . . . on such a global scale, and with such global visibility" (CGG 1995: 12).

This concept of globalization and a shrinking world is not without its critics. Some skeptics argue that while interdependence and technological advancement have increased in some parts of the world, this is not true in a vast majority of the South. (The terms *the South*, *the developing world*, *the less developed countries*, and *the third world* are used interchangeably throughout this book. They refer to the poorer countries, in contrast to the United States, Canada, Western Europe, Japan, Australia, and New Zealand, which are referred to as *the North*, *the more developed economies*, *the advanced industrial economies*, and *the first world*.) "'Global' is not 'universal'" (Mowlana 1995: 42). Although a small number of people in the South may have access to much of the new technology and truly live in the "global village," the large majority of the population in these countries does not. In most African countries there are fewer than four televisions for every 100 people (UNDP 1996). There are fewer phone lines in sub-Saharan Africa than there are in Manhattan (Redfern 1995), and "of the 600 million telephones in the world, 450 million of them are located in nine countries" (Toffler and Toffler 1991: 58).

Even those in the South that have access to television or radio are at a disadvantage. The globalization of communication in the less developed countries typically is a one-way proposition: the people do not control any of the information; they only receive it. It is also true that worldwide the ability to control or generate broadcasts rests in the hands of a tiny minority.

While lack of financial resources is an important impediment to globalization, there are other obstacles. Paradoxically, Benjamin Barber, who argues that we are experiencing global integration via "McDonaldization," asserts we are at the same time experiencing global disintegration. The breakup of the Soviet Union and Yugoslavia, as well as the great number of other ethnic and national conflicts (many of which are discussed in Chapter 3), are cited as evidence of forces countering globalization. Many subnational groups (groups within nations) desire to govern themselves; others see threats to their religious values and identity and therefore reject the secular nature of globalization. As a result, globalization has produced not uniformity, but a yearning for a return to non-secular values. Today, there is a rebirth of revitalized fundamentalism in all the world's major religions, whether Islam, Christianity, Judaism, Shintoism, or Confucianism. At the same time the global homogeneity has reached the airwaves, these religious tenets have reemerged as defining identities. (Mowlana 1995)

None of these criticisms mean that globalization, as we have defined it, is not occurring to some extent; they do, however, provide an important caution against overstating or making broad generalizations about the effects of globalization.

■ IS GLOBALIZATION GOOD OR BAD?

There are some aspects of globalization that most will agree are good (for example, the spread of medical technology) or bad (for example, increased global trade in illegal drugs). But other aspects are more complex.

The first column of Table 1.1 identifies three areas that are affected by globalization: politics, economics, and culture. A key aspect of political globalization is the weakened ability of the state to control both what crosses its borders and what goes on inside them. In other words, globalization can reduce the state's *sovereignty* (the state's ability to govern matters within its borders). This can be viewed as good, because undemocratic governments are finding it increasingly difficult to control the flow of information to and from prodemocracy groups. Satellite dishes and electronic mail are two examples of technology that have eroded state sovereignty. But decreased state sovereignty also means that the state has difficulty controlling the influx of illegal drugs, nuclear materials, unwanted immigrants, and terrorists.

In the realm of economics, increased globalization has given consumers more choices. Also, multinational corporations are creating jobs in poor areas where people never before had such opportunities. Some critics reject these points, arguing that increased foreign investment and trade benefit only a

Table 1.1 Advantages and Disadvantages of Globalization

Realm of Globalization	Advantages	Disadvantages
Political	Weakens power of authoritarian governments	Unwanted external influences are difficult to control
Economic	Jobs, capital, more choices	Exploitative; only benefits a few
Cultural	Offers exposure to other cultures	Cultural imperialism

small group of wealthy individuals and that, as a result, the gap between rich and poor grows both within countries and between countries. Related to this is the argument that many good-paying, blue-collar jobs are moving from the North to the poor countries of Latin America, Africa, and Asia.

At the cultural level, those who view increased cultural contact as positive say that it gives people more opportunities to learn about (and purchase goods from) other cultures. But critics of cultural globalization argue that the wealthy countries are guilty of cultural imperialism—that their multibillion-dollar advertising budgets are destroying the cultures of non-Western areas, as illustrated by Avon's aggressive sales strategy in the Amazon region (Byrd 1994).

The degree to which cultural values can be "exported" is the subject of some debate. Samuel Huntington argues that "drinking Coca-Cola does not make Russians think like Americans any more than eating sushi makes Americans think like Japanese. Throughout human history, fads and material goods have spread from one society to another without significantly altering the basic culture of the recipient society" (Huntington 1996: 28–29). Similarly, others argue that globalization brings only superficial change. "McDonald's may be in nearly every country, but in Japan, sushi is served alongside hamburgers. In many countries, hamburgers are not even on the menu" (Mowlana 1995: 46).

It is left to the reader to determine whether globalization is having a positive or negative effect on the issues discussed in this book. Is globalization enhancing our capability to deal with a particular issue? Or is it making it more difficult? It is left to the reader to determine whether globalization is having a positive or negative effect on the issues discussed in this book. Is globalization enhancing our capability to deal with a particular issue? Or is it making it more difficult? Of course, each individual's perspective will be influenced by whether he or she evaluates these issues based on self-interest, national interest, religious views, or from a global humanitarian viewpoint.

■ INTERCONNECTEDNESS AMONG ISSUES

As mentioned above, a primary purpose of this book is to explore how the issues introduced in the various chapters are interconnected. Table 1.2 is designed to illustrate this notion of linked issues. Each cell in the table represents the interaction of an issue in the first column with an issue in the top row. For example, Cell 2 (C2) should be read as follows: conflict (see the left column) can lead to negative consequences in the international economy (see the top row) as a result of war disrupting the free flow of goods between two countries or within an entire region.

Of course, when two global issues interact, the result is not necessarily negative. Cell 8 (C8) shows that an increase in a country's gross national product (GNP, or total of goods and services produced by a country's citizens in a given year) can mean a decline in its poverty rate. But also note that a possible linkage will not always occur: as Chapter 8 points out, an increase in GNP does not always lead to a decline in poverty.

Table 1.2 does not cover all possible linkages but points out a few basic ones. Also, the table understates the multiple nature of the linkages. For instance, the fall in poverty rates suggested in C8 would affect the environment, which in turn would affect international economic issues like trade, which in turn would affect poverty, and so on. Thus, each variable in Table 1.2 not only has multiple consequences, but also creates a ripple effect.

Table 1.2 Connections Between Global Issues

	Conflict	International Economics	Poverty	Population/ Migration	Environment
Conflict	C1 X	C2 war → disruption of trade patterns	C3 war → destruction of food crops	C4 conflict → emigration	C5 nuclear war → environmental damage
International Economics	C6 trade disputes → trade wars	C7 X	C8 increase in GNP → decrease in poverty	C9 decrease in jobs → emigration	C10 increase in GNP → increased pollution
Poverty	C11 increase in poverty → conflict	C12 poverty increases → more foreign investment sought	C13 X	C14 decrease in poverty → less emigration	C15 poverty → environmental destruction
Population/ Migration	C16 illegal immigrants → domestic conflict	C17 migrant labor → increase in low-wage jobs	C18 population increase → increase in number of poor	C19 X	C20 increase in population → strain on natural resources
Environment	C21 scarce resources → conflict	C22 abundant natural resources → wealth via exports	C23 unsustainable use of environment → poverty	C24 unsustainable development → emigration	C25 X

■ OUTLINE OF THE BOOK

This book has been organized into four parts. The first, which focuses on conflict and security issues, considers some of the primary sources of conflict and some of the many approaches to establishing and maintaining peace. Part 2 concentrates on economic issues ranging from international trade and investment to one of the major concerns that confronts the global economy—poverty. Part 3 deals with issues that, although not confined to, tend to plague the poorer countries. And Part 4 focuses on environmental issues and cooperative attempts to solve them. A concluding chapter discusses possible future world orders, sources of hope and challenges that face us in the coming century, and things individuals can do to have a positive impact on global problems.

■ QUESTIONS

1. What examples of globalization can you identify in your life?

2. Do you think globalization will continue to increase? If so, in what areas?

3. Do you think globalization has more positive attributes or more negative attributes?

4. Can you think of additional examples that could be included in Table 1.2?

■ SUGGESTED READINGS

Barber, Benjamin R. (1996) *Jihad vs. McWorld*. New York: Ballantine Books.
Bulletin of Atomic Scientists (1995) 51, no. 4 (July-August).
Hauchler, Ingomar, and Paul M. Kennedy, eds. (1994) *Global Trends: The World Almanac of Development and Peace*. New York: Continuum.
http://www.monde-diplomatique.fr/md/dossiers/ft/
Huntington, Samuel P. (1996) "The West: Unique, Not Universal," *Foreign Affairs* 75, no. 6.
Iyer, Pico (1993) "The Global Village Finally Arrives," *Time* 21, no. 142 (special issue).
King, Alexander, and Bertrand Schneider (1991) *The First Global Revolution*. New York: Pantheon Books.
New Perspectives Quarterly (1995) 12, no. 4.
United Nations Development Programme (1996) *Human Development Report*. New York: Oxford University Press.

Part 1

Conflict and Security

The Global Challenge of
Weapons Proliferation

Jeffrey S. Lantis

The proliferation of weapons is one of the most serious challenges to international security today. Arms races, regional competition, and the spread of weapons technology to other countries are all important dimensions of the proliferation challenge that could contribute to long-term global instability.

Proliferation is best understood as the rapid increase in the number and destructive capability of armaments. Evidence of the impact of proliferation on world affairs can be seen in the arms race between Germany and Great Britain that helped to spark World War I; the nuclear arms race between the superpowers, the United States and the Soviet Union, that brought us to the brink of a World War III; and the clandestine arms buildup in Iraq that helped it fight the Gulf War.

It is important to remember, however, that proliferation is not simply a problem for politicians and military leaders. When governments choose to use weapons in conflict in the twentieth century, they are exposing both soldiers and civilians to danger. In fact, the proliferation of weapons has contributed to higher civilian casualties and greater destruction this century than in the previous four centuries combined (Small and Singer 1982). When governments devote funds to build up large armies and weapons of mass destruction, they are also choosing to divert funds from other programs like education and health care. Clearly, citizens of the world experience these direct and indirect effects of proliferation every day.

■ TYPES OF PROLIFERATION

This chapter examines four different types of weapons proliferation. As illustrated in Table 2.1, there are two broad categories to consider: vertical

Table 2.1   The Proliferation Matrix

	Vertical Proliferation	Horizontal Proliferation
Conventional Weapons	Type I	Type II
Weapons of Mass Destruction	Type III	Type IV

versus horizontal proliferation; and conventional weapons versus weapons of mass destruction. Vertical proliferation is the buildup of armaments in one country. Horizontal proliferation is defined as the spread of weapons or weapons technology across country borders. Conventional weapons are those systems that make up the vast majority of all military arsenals—including most guns, tanks, planes, and ships. Weapons of mass destruction are those "special" weapons that have a devastating effect even when used in small numbers and kill more indiscriminantly than conventional weapons; they include nuclear, chemical, and biological systems.

■ Type I: Vertical Proliferation of Conventional Weapons

The buildup of conventional weapons arsenals in many countries is the oldest form of proliferation in human civilization and represents the foundation of the proliferation threat. At first glance, one might view this category of proliferation as the least threatening or most benign of all forms. Vertical conventional proliferation, however, can be a threat to international stability for at least two major reasons. First, arms buildups provide more weaponry for governments and groups to engage in more conflicts. At the same time, conventional weapons have become more sophisticated—from breech-loading rifles to precision-guided munitions—and more destructive—from mortar shells to multiple-launch rocket systems. Vertical conventional proliferation in an unregulated world market may provide determined leaders with enough incentive to order aggressive action and to actually spark conflicts. A second important danger of conventional arms buildups in one country is the social cost, which often includes serious reductions in social welfare spending by governments for citizens who can ill afford such deprivations.

More weapons means more violence. Government programs to build up conventional armaments ensure that there are more weapons available for countries to engage in more conflict. Some experts believe that the simple availability of weapons systems and the development of military strategy increases the chances that a country will engage in conflict. They have argued that advances in conventional weaponry and offensive military strategies were contributing factors to the outbreaks of numerous conflicts, including both world wars and the Vietnam War. In this context, arms buildups are seen as one potential cause of war in the international system (Sagan 1986; Sivard 1991).

In traditional forms, conventional arms buildups focus on weapons systems that are considered to be most effective for the times. In the period leading up to World War I, Germany and Great Britain engaged in a race to build the most powerful and awesome warships. In the period leading up to World War II, Adolf Hitler ordered research and development of rudimentary surface-to-surface missiles and jet aircraft as a way to gain military advantage. During the Cold War, President Ronald Reagan called for the creation of a 600-ship U.S. naval fleet, with an emphasis on strong aircraft carrier battle groups and advanced submarines. More recently, attention has turned to the latest technology of warfare, including stealth planes and ships, remote-controlled surveillance aircraft, antisatellite weapons, and computer technology that would give mobility and advantage to the fighting forces of the twenty-first century.

The relationship between arms buildups and the likelihood of conflict is multiplied by the fact that conventional weapons have become more sophisticated and destructive over the years. "Smart" conventional bombs and precision-guided munitions have improved both accuracy and the capability to do the kind of damage intended by the attacker. The increase in destructive capacity of conventional weapons such as fuel-air explosives and the faster and more accurate M1A1 tank also poses a greater threat to soldiers and civilians.

Finally, it is important to remember that conventional arms have been used repeatedly in conflict over the past fifty years. From landmines to fighter jets, conventional weapons have been blamed for roughly 50 million deaths around the globe since 1945. Individuals, groups, and governments have all built and used conventional weapons to achieve their goals.

The social costs of arms buildups. In the late twentieth century, many governments have built sizable conventional arsenals. The average level of U.S. government defense expenditures has topped $250 billion annually in the past two decades, with the majority of these funds going to support high levels of conventional weaponry and troops. In 1994, the Clinton administration sought a total defense budget request of $263.7 billion, which supported an active-duty military strength of 1,525,700 soldiers and a force structure composed of ten army divisions, twelve navy aircraft carrier task groups, and thirteen air force combat wings (Aspin 1994). Another way to interpret defense spending in 1994 is to say that the U.S. government spent about $1,000 on defense for every citizen. Or in relative terms, U.S. defense expenditures in 1994 were more than four times that of its nearest potential competitor, Russia (USG 1995). Figure 2.1 illustrates the broader context of changing levels of global defense expenditures.

Figure 2.1   Global Defense Expeditures, 1987–2000

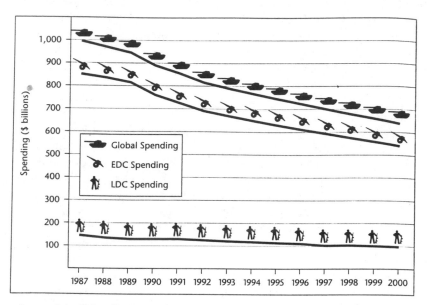

Source: John T. Rourke, *International Politics on the World Stage* (Guilford, CT: Dushkin Publishing Group/Brown and Benchmark Publishers, 1995), p. 270.

Note: Expenditures are calculated in billions of 1991 U.S. dollars. Dollar amounts for all years are not shown.

While Figure 2.1 shows a decline in global defense spending, it is clear that countries continue to spend hundreds of billions of dollars every year on the military. This has led many critics to charge that there are dangerous social costs in the trade-off between "guns and butter," and the end of the Cold War has drawn new attention to this difficult balance between military and social spending.

In the 1990s, the United States is first in the world in terms of military spending but ranks relatively low against other countries on various social indicators. International relations experts Charles Kegley and Eugene Wittkopf (1997) point out that the United States is only fourth in the world in terms of literacy rates, ninth in per capita public expenditures for education, thirteenth in average scores of students on science and math tests, twenty-first in infant mortality, and twenty-fifth in percentage of population with access to sanitation. A related study found that when military expenditures rose in developing countries during the past few decades, the rate of economic growth declined and government debt increased (Nincic 1982). Kegley and Wittkopf conclude that it is a sad truth that many countries have become more concerned with defending their citizens from foreign attack in the twentieth century than they are with protecting them from social, educational, and health insecurities at home.

■ Type II: Horizontal Proliferation of Conventional Weapons

A second category of proliferation is the horizontal spread of conventional weapons and related technology across country borders. The main route of the spread of conventional weaponry is through legitimate arms sales. But the conventional arms trade has become quite lucrative and many experts are concerned that the imperative of the bottom dollar is driving us more rapidly toward global instability.

Arms dealers. The conventional arms trade has become a very big business, and the great powers—the United States, France, Russia, Great Britain, and China—are major dealers of conventional arms. In 1987, the Soviet Union was at the top of the arms trade, dominating the market with 46 percent of all sales. But as Soviet and (later) Russian sales levels plummeted, the United States quickly emerged as the new leader. In 1994, the Congressional Research Service reported that the United States had captured true dominance in the global arms market with 47 percent of all sales. Seven years after the end of the Cold War, it was the United States government and defense industries that were marketing advanced conventional weapons around the world. In the wake of the successful demonstration of the effectiveness of U.S. weapons through the Gulf War, military contractors made large shipments of F-15 fighter aircraft to Saudi Arabia, sold hundreds of M1A1 tanks to Kuwait, and finalized many other similar deals. In the 1990s, conventional arms sales have earned U.S.-based defense contractors about $8 billion annually, and in 1995 U.S. companies actually produced more fighter jets for export than for purchase by the U.S. military (*Boston Globe* 1996).

Arms customers. The sales figures for the top arms merchants are significant in themselves, but it is also important to identify key customers. Generally speaking, U.S. defense contractors have sold a great deal of hardware to allies. U.S. arms deliveries to Israel from 1984 to 1993 were an estimated $9.5 billion, which included 450 armored combat vehicles, fifty used F-16s, and twenty-five new F-15s. In southern Europe, Greece purchased $4 billion worth of U.S. arms from 1991 to 1994. In Asia, allies, including Japan, South Korea, and Taiwan, have purchased large numbers of U.S. conventional weapons systems (Hartung 1995).

Arms sales are not always made to countries considered traditional allies, however. From 1984 to 1989, the People's Republic of China purchased some $424 million in U.S. weapons, and these arms deals were stopped only after the Tiananmen Square massacre of prodemocracy activists in the summer of 1989. Through legitimate means, Iraqi president Saddam Hussein purchased a massive conventional arsenal on the international arms market. In 1990, estimates of the arsenal included a total of 5,500 tanks, 4,000 heavy artillery, 7,500 armored personnel carriers, and 700 planes. Arms sales to Iraq by friends and allies came back to haunt the United States, however, in the Gulf War, and the sale of conventional weapons raises real concern about the potential for "deadly returns" on U.S. investments (Laurance 1992).

■ **Type III: Vertical Proliferation of**
Weapons of Mass Destruction

The vertical proliferation of weapons of mass destruction (WMD) is another serious threat to international security. There are several important dimensions of this problem, including the range and variety of modern WMD systems, incentives for states to build nuclear weapons, and the patterns of vertical WMD proliferation.

Types of weapons of mass destruction. There are three different types of weapons of mass destruction in existence today: nuclear, biological, and chemical. These are often examined as a group, but it is important to note that their effects and their potential military applications are quite different.

Nuclear fission was discovered in 1938, and scientists like Albert Einstein soon called on governments to sponsor an exploration of its potential. Atomic weapons were first developed by the U.S. government through the five-year, $2 billion secret research program during World War II known as the Manhattan Project. On August 6, 1945, the United States dropped a 12.5-kiloton atomic bomb on Hiroshima, Japan. This weapon produced an explosive blast equal to that of 12,500 tons of conventional high explosives (like TNT) and caused high-pressure waves, flying debris, extreme heat, and radioactive fallout. A second bomb was dropped on Nagasaki on August 9, 1945, and the Japanese government surrendered one day later (Schlesinger 1993).

The use of atomic bombs to end World War II in 1945 was actually the beginning of a very dangerous period of spiraling arms races between the United States and the Soviet Union. The Soviet regime immediately stepped up its atomic research and development program. In 1949, they detonated their first atomic test device and joined the nuclear club. By the 1980s, the Soviet Union had accumulated an estimated 27,000 nuclear weapons in its stockpile. Both superpowers also put an emphasis on diversification of their weapons systems. The symbolic centerpiece of each side's nuclear arsenals was their land-based inter-continental ballistic missiles (ICBMs), which as their name implies can be launched from one continent to another. But each side had also deployed nuclear weapons on submarines; in long-range bombers; as warheads on short-range, battle-field missile systems; and even in artillery shells and landmines.

Chemical weapons and biological weapons. Chemical weapons, another class of weapons of mass destruction, work by spreading poisons that can incapacitate, injure, or kill through their toxic effects on the body. These clearly antipersonnel weapons can be lethal when vaporized and inhaled in very small amounts or when absorbed into the bloodstream through skin contact. Examples of chemical weapons range from tear gas used by riot police to disperse crowds to nerve agents such as Sarin (recently used by a radical religious cult in Japan to terrorize civilians in Tokyo).

Many governments and independent groups have funded chemical weapons research and development programs in the twentieth century. In fact, chemical weapons are relatively simple and cheap to produce com-pared with other classes of WMDs. Any group with access to basic chem-ical manufacturing plants or petrochemical facilities can develop variants of commonly used, safe chemicals to create dangerous weapons of mass destruction. The first recorded use of chemical weapons in warfare oc-curred in the fifth century B.C.E. when Athenian soldiers poisoned their enemy's water supply with a chemical to make them sick. The last known wide-scale use occurred during the Iran-Iraq War (1980–1988), where an estimated 13,000 soldiers were killed by chemical weapons (McNaugher 1990).

As dangerous as chemical agents can be, biological agents are actually much more lethal and destructive. Biological agents are basically disease-causing microorganisms such as bacteria, viruses, or fungi that can be de-ployed to cause massive infections that incapacitate or kill the intended target after an incubation period. A more lethal derivative of biological weapons—toxins—can cause incapacitation or death within minutes or hours. Examples include anthrax, a disease-causing bacteria that contains as many as 10 million lethal doses per gram.

Like chemical agents, biological and toxic weapons are relatively easy to construct and have a high potential lethality rate. Any government or group with access to basic pharmaceutical manufacturing facilities or bio-logical research facilities can develop biological weapons. And, like the other classes of WMD systems, information about the construction of such systems is available in the open scientific literature and on the Internet.

Why build WMD systems? There are two basic reasons why countries build weapons of mass destruction: security and prestige. First, many gov-ernment leaders genuinely believe that their state security is at risk with-out such systems. The standoff between India and Pakistan is an example. After years of rivalry and border skirmishes between the countries, India began a secret program to construct an atomic device that might swing the balance of regional power in their favor. In 1974, the Indian government detonated what it termed a "peaceful nuclear explosion"—signaling their capabilities to the world and threatening Pakistani security. For the next twenty-five years, both Pakistan and India secretly developed nuclear weapons in a regional arms race. In May 1998, the Indian government detonated five more underground nuclear explosions and the Pakistani government responded to the perceived threat with six nuclear explosions of its own. At this writing, the two governments have acknowledged their nuclear capabilities and have warned that they may place nuclear warheads on missiles targeted against one another. Another example comes from the Middle East, where Israel is suspected of having developed dozens of nuclear devices for potential use in their own defense. There are now reports that the Israeli government threatened to use these systems against Iraq during the Gulf War if Israel came under chemical or biological weapons attack (Schlesinger 1993).

Second, some governments have undertaken WMD research and development programs for reasons of prestige, national pride, or influence. It became clear to some during the Cold War that the possession of WMD systems lent a certain level of prestige, power, and even influence to state affairs. At a minimum, the possession of WMD systems—or a spirited drive to attain them—would gain attention for a country or leader. North Korea's drive to build a nuclear device based on an advanced uranium enrichment process drew the attention of the United States and other Western powers in the early 1990s. After extensive negotiations, North Korea was offered new nuclear energy reactors in exchange for a promise not to divert nuclear material for a bomb program.

Other government leaders pursue the development of WMD arsenals because they believe that it will help them gain political dominance in their region of the world. To illustrate this dynamic, Gerald Steinberg (1994), an expert on proliferation, relates the story of clandestine Iraqi government efforts to develop a WMD arsenal. Iraqi leader Saddam Hussein ordered the creation of a secret WMD research and development program and began to acquire nuclear technology and materials from France, Germany, the United States, and other countries in the late 1970s. While research scientists in the program worked on uranium enrichment, Saddam Hussein worked to improve his political profile in the region and to improve relations with key Arab states. Meanwhile, the Israeli government tried to stop the clandestine nuclear program by carrying out a devastating air strike against Iraq's nuclear research reactor at Osiraq in 1981. But the determined Iraqi drive for regional influence was really only stopped by the efforts of the U.S.-led international coalition in the Gulf War and the dispatch of a United Nations (UN) special commission to investigate and dismantle the Iraqi WMD development program. Broadly speaking, Iraqi proliferation efforts were part of a larger scheme to gain prestige, power, and influence in the Middle East.

■ Type IV: Horizontal Proliferation
of Weapons of Mass Destruction

The horizontal proliferation of WMD systems represents the final dimension of this challenge to international peace and stability. In fact, the spread of these weapons and vital technology across state borders is often viewed as the most serious of all proliferation threats.

Nuclear arsenals. The massive buildup of nuclear arsenals by the superpowers was not the only game in town during the Cold War. In fact, while the Soviet Union and United States were stockpiling their weapons, several other states were working to join the nuclear club through both open and clandestine routes.

Today the United States, Russia, France, Great Britain, and China all openly acknowledge possessing stocks of nuclear weapons. At the height of the Cold War, the United States and Soviet Union supported key allies by secretly authorizing the transfer of sensitive nuclear weapons technology to other research and development programs. In 1952, Great Britain successfully tested an atomic device and eventually built a nuclear arsenal that today numbers about 200 weapons. France officially joined the nuclear club in 1960 and built a somewhat larger nuclear arsenal of an estimated 420 weapons. The People's Republic of China detonated its first atomic device in 1964 and built an arsenal of about 300 nuclear weapons during the Cold War (McGwire 1994).

The controlled spread of nuclear weapons and weapons technology from the superpowers to key allies was not the only route by which countries might obtain valuable information and materials. Several less developed countries began secret atomic weapons research and development projects after World War II. As noted earlier, states like India, Pakistan, and Israel have pursued clandestine WMD programs because of concerns about security and prestige. In some cases, these efforts were facilitated by covert shipments of material and technology from the great powers, but research and development of WMD systems was also aided by the availability of information in the open scientific literature (and by the resourcefulness of scientists and engineers).

When the Indian government detonated its first nuclear explosion in 1974, it symbolically ended the monopoly on nuclear systems held by the great powers. India actually obtained nuclear material for their bomb by diverting it from a Canadian-supplied nuclear energy reactor that had key components originally made in the United States. Most experts believe that India now possesses a significant stockpile of about fifty unassembled nuclear weapons. The 1974 Indian detonation was, of course, a catalyst for the Pakistani government to step up its research and development program, and today most experts believe that Pakistan has an arsenal of dozens of weapons that could be quickly assembled for use. The test explosions sponsored by both governments in 1998 were further evidence of their capabilities.

Finally, Israel may possess as many as 100 nuclear weapons. The Israeli nuclear program was a derivative of research and development projects in the United States and, ironically, the Soviet Union. Like India, the Israeli government proved to be quite resourceful in adapting existing technologies to construct their arsenal (Forsberg, Driscoll, Webb, and Dean 1995).

Finally, there are former nuclear states that have made political decisions to dismantle their weapons. Included in this group are South Africa and three former Soviet republics: Belarus, Kazakhstan, and Ukraine. The South African government recently admitted that it had constructed several nuclear devices for self-defense in the 1960s. But the government decided to destroy these weapons at the end of the 1980s—unilaterally removing themselves from the nuclear club. The three former Soviet republics each had strategic nuclear weapons stationed on their territory after the breakup of the Soviet Union. Soon after gaining their independence, however, the three republics agreed to become nonnuclear weapons states under the Treaty on the Non-Proliferation of Nuclear Weapons (or Non-Proliferation Treaty, NPT). In 1992, they signed on to the Lisbon Protocol to the Strategic Arms Reduction Treaty (START), agreeing to transfer all the nuclear warheads on their territory to Russia in exchange for economic assistance from the United States (McGwire 1994).

The spread of chemical and biological weapons. Over 100 countries have the basic capability to develop chemical or biological weapons, and at least twenty countries have already done so. Figure 2.2 illustrates the range of actors involved in the proliferation of weapons of mass destruction.

WMD terrorism. The horizontal proliferation of WMD systems raises another concern about international security in the late twentieth century: the possibility that nuclear, chemical, or biological weapons systems may be used in terrorist attacks. Unfortunately, this fear was realized in 1995 when a Japanese religious cult released Sarin nerve gas into the Tokyo subway system, killing ten civilians and injuring more than 5,000. This attack confirmed the growing concerns about the monitoring and control of terrorist groups and their weapons capabilities.

Many experts believe that the horizontal spread of nuclear weapons, material, and know-how has dramatically increased the likelihood that a group or state will attempt an act of nuclear terrorism in the future. This is of particular concern given the chaos and instability surrounding the nuclear arsenal of the former Soviet Union, and there have been numerous reports in the past few years of attempts to buy or steal nuclear warheads in that region. In a 1994 feature story in the *Atlantic Monthly*, Seymour Hersh reports that in January 1991, armed Azeri rebels in Azerbaijan penetrated a Soviet base on which tactical nuclear weapons were stored and gained physical access to a nuclear warhead for a short period before being

Figure 2.2   Countries Suspected of Having Programs to Develop Weapons of
                Mass Destruction

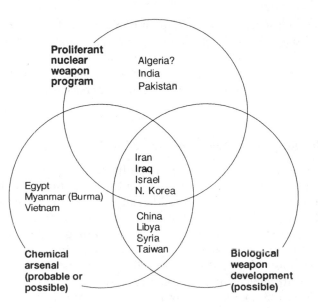

Source: Randall Forsberg, William Driscoll, Gregory Webb, and Jonathan Dean, eds.,
*The Nonproliferation Primer: Preventing the Spread of Nuclear, Chemical, and Biological
Weapons* (Cambridge, MA: Institute for Defense and Disarmament Studies, 1995). Used by
permission of the editors.
  Notes: Iraqi programs reversed by UN.
  China is an acknowledged nuclear weapon state.

ejected by Soviet troops. Later that year, a nuclear expert from Greenpeace
actually arranged a secret deal to purchase a Soviet nuclear warhead for
$250,000 from a group of disgruntled Russian soldiers.

    The emergence of fifteen newly independent states in the region with
very porous borders also increases the likelihood that nuclear materials and
know-how will be smuggled out of the country. In Germany, for example,
more than 100 arrests have been made in connection with attempts to smug-
gle nuclear materials out of the former Soviet Union in the 1990s. In October
1992, two containers of radioactive material were discovered by the police in
Frankfurt containing amounts of cesium and strontium misappropriated from
scientific or medical establishments in the former Soviet Union. Meanwhile,
nuclear scientists in the former Soviet republics (who currently earn an average monthly salary of $30 in defunct research facilities) are being lured to
less developed states like Iran to work in budding nuclear research programs
by the promise of high wages and social status. All these factors suggest that
the threat of nuclear terrorism has indeed increased in the post–Cold War era
(Forsberg, Driscoll, Webb, and Dean 1995).

■ GLOBAL SOLUTIONS: NONPROLIFERATION INITIATIVES

Proliferation is a very complex and multilayered challenge to international security. Many world leaders and experts have recognized this threat and have begun to address the proliferation challenge through a series of regional and global nonproliferation initiatives. The scope, number, and momentum of these initiatives have steadily increased over the past few decades.

■ Global Responses—the Nonproliferation Regime

In some ways, the global nuclear nonproliferation movement began even before the first use of atomic weapons in 1945. Politicians, military leaders, and scientists involved in the Manhattan Project recognized from the outset that such weapons were somehow special and more dangerous than other systems. President Truman, who had ordered the use of atomic bombs over Hiroshima and Nagasaki, authorized his ambassador to the United Nations, Bernard Baruch, to deliver a proposal to the organization calling for all nuclear materials and technology to be placed under UN oversight. While the plan did not receive widespread support, it demonstrated a first step toward global consideration of proliferation problems and set the stage for later progress on the issue.

Ten years later, in 1956, member states of the UN agreed to create the International Atomic Energy Agency (IAEA). The IAEA's primary mission was to serve as a watchdog and inspection organization to promote the peaceful uses of nuclear technology and stop adaptation for military uses. This agency soon became an integral part of the global nuclear nonproliferation regime.

In the 1960s, world leaders agreed to new initiatives, including the Partial Nuclear Test Ban Treaty. This agreement banned nuclear tests in the atmosphere, in outer space, and under water, and it was originally signed by leaders from the Soviet Union, Great Britain, and the United States. In 1967, the Treaty for the Prohibition of Nuclear Weapons in Latin America created the first large nuclear-free zone. Signatories to this treaty pledged to use nuclear facilities only for peaceful purposes. Most important, the treaty outlawed the testing or acquisition of nuclear weapons in the region and even precluded third parties from bringing weapons to the region (Davis 1991).

■ The Nuclear Non-Proliferation Treaty

The NPT represented one of the most significant advances in the development of the global nonproliferation regime. The NPT was an agreement to halt the spread of nuclear weapons beyond the five declared nuclear powers. Specifically, the treaty had ambitious goals for both vertical and horizontal proliferation. Article I of the treaty dictated that no nuclear weapons state (defined by the treaty as a state that detonated a nuclear explosive prior to 1967) would transfer "directly or indirectly" nuclear weapons, explosive devices, or control over these weapons to another party. Article II stipulated that no nonnuclear weapons state could receive, manufacture, obtain assistance for manufacturing, or otherwise try to acquire nuclear weaponry. Article VI obligated all nuclear states to pursue disarmament, and other sections of the treaty required all nonnuclear weapons states to accept full-scope nuclear safeguards established and monitored by the IAEA (Roberts 1995).

The NPT represented the crowning achievement of global nonproliferation efforts during the Cold War. After careful and extensive negotiations, the treaty was signed by all the acknowledged nuclear states and 168 others. The NPT went into effect in 1970. In 1995, world leaders gathered at a special Review and Extension Conference to reconsider and evaluate the effectiveness of the treaty. After some debate, they declared that the treaty would "continue in force indefinitely" as a guarantor of international peace and security. Signatories also agreed to continue to pursue the ultimate goals of eliminating nuclear weapons and completing a treaty on general disarmament under strict and effective international control.

■ Related Nonproliferation Initiatives

Several other significant agreements have followed in the spirit of the NPT. The Biological Weapons Convention (BWC) of 1972 was the first formal effort to gain some control over the world's deadly biological arsenal. More than 100 signatories agreed to ban "the development, production and stockpiling of microbial or other biological agents." The convention, however, did not sanction nonsignatories and did not preclude research on biological weapons (Davis 1991).

In the same spirit, world leaders drafted a Chemical Weapons Convention (CWC) that was opened for signature in January 1993, after years of intensive negotiations. In many ways the framework of the CWC was similar to the structure of the NPT and the BWC. It committed all signatories to eliminate their stockpiles of chemical weaponry and to halt all development efforts. In addition, it included a set of verification procedures somewhat more stringent than those under the NPT. These procedures supported the rights of a new CWC Inspectorate to conduct rigorous investigations and surprise "challenge inspections" of suspected chemical weapons programs in signatory states. Ratification by the legislatures of sixty-five countries in the system was required for the CWC to come into force. After a great deal of debate about the implications of the treaty for

U.S. national security and sovereignty, the U.S. Congress finally ratified the agreement in April 1997.

The Comprehensive Nuclear Test Ban Treaty (CTBT), another nonproliferation initiative, was opened for signature in 1996. A large majority of UN member states voted to support the CTBT (a treaty that would eliminate all actual nuclear testing), and world leaders began to sign the treaty in the fall of 1996. To become international law, the treaty requires the signature of all forty-four countries known to possess nuclear reactors. By early 1998, representatives of the five declared nuclear powers had all signed the CTBT, but both India and Pakistan refused to do so. India has claimed that it wants the CTBT to be stronger in order to force nuclear states' compliance with Article VI of the NPT. Meanwhile, the Pakistani government has stated that it would not sign the CTBT without Indian cooperation. Both countries' nuclear tests of May 1998 underscored their resistance to this latest initiative of the nonproliferation regime. Several other states also remain reluctant to sign the treaty, including Libya, Cuba, and Syria (Crossette 1996a).

■ Controlling Weapons at the Point of Supply

There are also important concerns about the implications of the spread of WMD technology around the world. The NPT, for example, did not prevent states from exporting other types of materials that could potentially be adapted for use in the development of WMD programs. World leaders have tried to address the problem of weapons technology transfers for several decades. In 1976, major supplier states—including the Soviet Union, Japan, France, the United States, Great Britain, West Germany, and Canada—agreed to establish a "trigger list" of items that could be sold to other countries only under IAEA safeguards. Representatives of these states met in London, and this "London Club" established and coordinated a supply control group. In the 1980s, supplier states established the ballistic Missile Technology Control Regime (MTCR), which prohibited the transfer of essential technology for this aspect of weapons programs.

Like the NPT, however, supply control efforts have had only a mixed rate of success. They helped limit missile development projects under way in South America and the Middle East but allowed some twenty countries to join the ballistic missile club. These supply controls did not prevent Iraq from making significant progress toward the development of nuclear weapons through the modification of civilian scientific technology that was adapted for military use. And they did not prevent Iraq from manufacturing and modifying the Scud-B missiles, which were used against Israel and Saudi Arabia during the Gulf War—and which were capable of carrying chemical warheads. Meanwhile, Pakistan developed its own ballistic missile, the Hatf, and acquired about thirty nuclear-capable medium-range

M-11 missiles from China (McNaugher 1990). Today the North Korean government is developing a long-range Taepo Dong missile that may someday have the potential to reach the United States.

▓ The U.S. Response: Counterproliferation Strategies

In the post–Cold War era, President Clinton identified proliferation as a key threat to U.S. national security and argued that the government needed a coherent policy to deal with new states that acquired nuclear, chemical, or biological weapons. Given the dangers of proliferation and new deterrent dynamics, President Clinton authorized a comprehensive Nuclear Posture Review in October 1993 to define and adapt nuclear weapons to the changing international security environment. Even before the completion of the review, the administration had already begun "vigorous counterproliferation and threat-reduction efforts" including (1) an improvement of intelligence monitoring of proliferation; (2) an enhancement of the United States' ability to destroy, seize, and disable nuclear, chemical, or biological systems; (3) the development of ballistic and cruise missile defense; and (4) better cooperation with friendly governments to improve export control measures.

In November 1994, President Clinton took counterproliferation efforts a step further by issuing Executive Order No. 12938, which declared proliferation a "national emergency." In the face of such an emergency, the administration pushed for the implementation of arms reduction agreements like the Strategic Arms Reduction Treaty II, which called for a reduction of Russian and U.S. nuclear warheads to a level of 3,500, and the Conventional Forces in Europe Treaty, which would reduce military hardware levels across Europe. The Clinton administration also supported a bipartisan Senate initiative known as the Nunn-Lugar Plan to promote and oversee weapons dismantling programs, demilitarization, and defense diversification. Each of these programs has contributed to a new U.S. security policy posture that recognizes the contemporary challenge of proliferation (Rathjens 1995).

■ CONCLUSION: PROSPECTS FOR THE FUTURE?

The proliferation of weapons is truly a major challenge to global security, but there are also reasons for optimism about the prospects for solutions.

One of the most important catalysts of global proliferation was the Cold War arms race between the superpowers. Today, many scholars and politicians are taking a new look at incentives for proliferation in the post–Cold War era, and some say that we may be headed toward a nuclear-free twenty-first century. Optimists argue that a global build-down in tensions—

a reverse proliferation—has occurred with the end of Cold War tensions. They cite the completion of START I in January 1991, by which the two superpowers pledged to reduce their arsenals to between 8,000 and 9,000 weapons, as evidence of this trend toward reverse proliferation. Further momentum was gained two years later when the United States and Russia agreed to sign START II, reducing nuclear arsenals to no more than 3,500 warheads each. And the Clinton administration has now begun preliminary negotiations with Russia on a START III accord that would lead to further, dramatic reductions in pursuit of a "minimum nuclear deterrent" relationship (Bundy, Crowe, and Drell 1993).

Meanwhile, the indefinite extension of the NPT and the establishment of the CTBT both suggest an emerging global consensus to stop nuclear proliferation. On the conventional weapons front, there is growing recognition that conventional arms transfers—even small arms—also represent a threat to international security. UN experts and government leaders have been discussing ways to increase the transparency of the conventional arms trade by making more information available on arms transfer policies and data. Efforts are under way for both general classes of weapons transfers to enforce trade regulations and closely monitor weapons transactions (Karp 1994).

Solutions to the proliferation challenge must go even deeper, however. As we look toward the twenty-first century, citizens of the world must agree to build on the momentum of recent progress by making moral and principled stands against proliferation. For instance, Oscar Arias, the 1987 Nobel Peace Prize winner, has recently called for a global agreement to stop arms sales to countries that have violated human rights. This certainly would be an important step in a global effort to address ethical and moral concerns about the development of certain classes of weapons such as landmines and chemical and biological weapons. A moral stand against proliferation in favor of economic development, health care, and education may pave the way toward real peace and justice in the twenty-first century.

■ QUESTIONS

1. In your opinion, which of the four types of proliferation represents the most serious threat to international security?

2. Is the proliferation of conventional weapons a challenge that can ever fully be met by the global community? Why or why not?

3. Is it possible that weapons proliferation could actually make the international system more stable in the twenty-first century? How might this occur?

4. What are some of the efforts that individual countries and international organizations have made to respond to the proliferation challenge? Which are most effective, and why?

5. What are some of the implications of the trade-off between expenditures on defense and social welfare programs?

6. Can countries afford to enjoy a "peace dividend" in the post–Cold War era by diverting large sums from defense expenditures to other needs? Can they afford not to?

7. In your opinion, should government leaders offer to pursue complete WMD disarmament? Why or why not?

■ SUGGESTED READINGS

Bailey, Kathleen C. (1993) *Strengthening Nuclear Non-Proliferation*. Boulder, CO: Westview Press.

Karp, Aaron (1994) "The Arms Trade Revolution: The Major Impact of Small Arms," *Washington Quarterly* 17 (Autumn).

Krause, Keith (1992) *Arms and the State: Patterns of Military Production and Trade*. Cambridge: Cambridge University Press.

Moodie, Michael (1995) "Beyond Proliferation: The Challenge of Technology Diffusion," *Washington Quarterly* 18 (Spring).

Quester, George H., and Victor A. Utgoff (1994) "Toward an International Nuclear Security Policy," *Washington Quarterly* 17 (Winter).

Sagan, Scott D., and Kenneth N. Waltz (1995) *The Spread of Nuclear Weapons: A Debate*. New York: W.W. Norton.

Nationalism

John K. Cox

Nationalism is a complicated and widespread phenomenon in modern life. At its most basic level, nationalism is something felt by an individual. It is a sense of belonging. This belonging links the individual to a group of people on the basis of certain shared characteristics. Most important among these are a common language, a common history, and common customs or cultural traditions (sometimes including religion). When this individual national feeling develops into group national identity, nationalism becomes political. The term *self-determination* is used to describe the perceived right of every nation, or people, to rule itself. Ideally, this means that the various countries, or states, of the world would become "nation-states" (independent countries composed of members of a single national group) if their populations have nationalist feelings. This is a very complicated procedure, however, since many great empires and countries of the nineteenth and twentieth centuries were, and still are, decidedly multi-ethnic or multinational. Thus, only a small fraction of today's countries are true nation-states.

■ THE HISTORY OF NATIONALISM

Napoleon Bonaparte, who ruled France from 1799 to 1815, is usually credited with introducing the modern concept of nationalism. It spread throughout Europe as a result of the Napoleonic Wars. This new national feeling went far beyond simple patriotism, which is the love of one's homeland. Patriotism has been a part of human behavior since the beginning of history. Usually considered noble, it most often found expression in military terms. But it was a rather narrow idea compared to today's nationalism, since it was limited to the religious duty to "die a good death," was bound up with feudal localism, or was restricted to one stratum of society, usually an upper class (Teich and Porter 1993: xviii).

Nationalism as we know it was a product of the French Revolution. In 1789, France was seized by massive protests and revolts, while many of its intellectuals were under the influence of the Enlightenment. The *ancien régime* (old government and social system) of France had run the country deep into debt. Important economic and demographic shifts had taken place with the growth of cities and industries and commerce. The classes that participated in such activities, such as skilled and unskilled workers, and merchants, were clamoring for more political power. But the system was still controlled by the Bourbon family, France's absolutist kings who claimed rule by divine right. The disturbances in France eventually brought about a democratic government. French nationalism was thus born in people's minds when their government truly became theirs for the first time. People, formerly known as subjects, became citizens in the new French democracy.

Most scholars who deal with nationalism—historians, political scientists, and sociologists—believe that the growth of nationalism is a fundamental aspect of modernization. Of course modernization involves more than self-determination; generally it involves industrialization, urbanization, increased literacy, and secularization. This was as true of European history in the nineteenth century as it is of the history of the decolonizing world—mostly Africa and Asia—in the twentieth. Therefore, the growth of nationalism involves two processes: its appearance in people's minds as a sense of loyalty and belonging; and its appearance as a political force, which ultimately works to create nation-states.

The transition to an industrial society often predisposes people to become more nationalistic, since the breakdown of traditional village and family structures leaves emotional and moral gaps in individual lives. Also, the centralization of government, which originated in the Middle Ages in the struggle of kings against recalcitrant nobles, was to tap into nationalism as a way to mobilize the population. Napoleon was the first to make use of this great power of the people by appealing to them with the symbolism and emotions of national unity and a national mission.

In the nineteenth century, people began to accept nationalism only gradually. Europe continued to be the main place where nationalism grew. In the many small states that made up the German cultural realm, for instance, nationalists and advocates of greater democracy joined forces. They were at once attracted to the power of nationalism and disturbed by France's use of it against them. Philosophers, publicists, and revolutionaries portrayed drives for national unity and independence in the best possible light; they said, in essence, that the diversity among nations was a blessing from God. Separating people into nations was thought to be a duty that would result in the maximum use of individuals' talents in the overall service of humanity. One of the leading idealistic nationalists was French historian Ernest Renan, who claimed that a nation was not built on ethnic or religious criteria, but on a "rich legacy of memories" and a "common will in the present"; in short, Renan said that a nation was a "spiritual principle" (Renan 1996: 52).

Skeptics and detractors of the movement, however, had very strong arguments against it. Leaders of Europe's many multinational states (Great Britain, Russia, and the Hapsburg and Ottoman empires), the nobility, many officials of the Catholic church, and Marxists, all for different reasons, opposed nationalism.

Despite this diverse and often intense opposition, the twentieth century began with what most people regard as the triumph—or running amok—of the national idea. The great powers of Europe, such as Germany, France, and Russia, became imperialistic and sought to expand their power at their neighbors' expense. They were filled with national pride, became aggressive, and organized themselves into massive alliance systems. A result was World War I, in which large numbers of troops combined with the propaganda of national glory and the vilification of the enemy to create a new level of battlefield fury and destructiveness. The Great War, as it is sometimes called, was the first total war involving weapons and tactics of mass destruction such as poison gas, tank assaults, and the bombardment of civilian population centers.

There is another important connection between World War I and nationalism: the realm of nation-states in Europe was greatly expanded as a result of the fighting. The old multinational empires of Europe collapsed. In their place arose a set of what diplomats endorsed at the time as nation-states, from Finland in the north to Turkey in the south. These included Poland, Estonia, Latvia, Lithuania, Hungary, and Austria, as well as two small, multiethnic confederations of mostly related peoples—Czechoslovakia and Yugoslavia.

■ TYPES OF NATIONALISM

One way to categorize types of nationalism is by their organizing concepts. What is a "nation"? Who belongs to it? And who is an outsider? In general, we may say that nationalism is broken down into two types. The first, and oldest, was initially associated with Western European or North American politics and with countries elsewhere that followed them. It is usually called "civic" or political nationalism and it is seen above all as a "legal-political concept," or as a "political configuration" (Bojtar 1988: 254). This type of nationalism is heir to the legacy of the French and American revolutions. In these revolutions, the growing middle class, or bourgeoisie, was carving out space in the political structure for itself. These revolutions are thus important milestones on the path to democracy, since they resulted in breaking the stranglehold on political power of the kings and aristocrats. Still, these middle classes were not interested in giving the vote immediately to the lower classes or to women. In theory, though, civic nationalism presupposes that citizenship and nationality are identical (Liebich, Warner, and Dragovic 1995: 186). The nation is a political population, united in its ideas and habits.

The other type of nationalism, "ethnic" nationalism, was originally associated with countries in Eastern and Central Europe. This nationalism is based on "ancestral association" (Bojtar 1988: 254) as compared to civic nationalism, which can embrace diverse people who live within shared borders. Ethnic nationalism requires a common culture, way of life, and above all a perceived sense of genetic links (as in a greatly extended family) to the members of the ethnic community. The word *ethnic* comes from the Greek word *ethnos* meaning a group of people united by their common birth or descent. It should be noted that all types of nationalism are in some way exclusionary. If nothing else, this is true because of the presence of borders and frontiers. But ethnic nationalism, due to its emphasis on the "blood line" or racial connections between citizens, is far more exclusionary than civic nationalism and pays less attention to political boundaries.

The historical differences between these types of nationalism are great and remain relevant to this day. The war in Bosnia-Herzegovina can be better understood by remembering that many Serbs and Croats adhere to the kind of exclusive nationalism of the second category (see Case Study One: Yugoslavia, p. 38). Why? Because the more inclusive civic nationalism of Western Europe was developing in the spirit of certain key turning points in European civilization, such as the Enlightenment and the growth of middle-class democracy. Western European nationalism arose in societies that were already modernizing, while the peoples of Eastern Europe were neither independent nor economically modern. In short, Eastern Europe became nationally conscious before it had experienced economic development, representative government, and political unity (or in many cases even independence from foreign rule). The result was a desire to alter the political boundaries to coincide with the national or cultural boundaries (Sugar and Lederer 1994: 10; Kohn 1965: 29–30).

The much-heralded civic nationalism can also be exclusive. For instance, the U.S. Constitution was designed in the 1780s to deny women and slaves the right to vote. It was only after the Civil War that African American men were officially given the right to vote (the Fifteeth Amendment), and in many states this right was not protected by meaningful enforcement of laws until the Civil Rights Acts of the 1960s, almost 100 years later. Women were denied the right to vote almost everywhere until the twentieth century; in the United States, this right was provided by the Nineteenth Amendment in 1920.

■ FUNCTIONS OF NATIONALISM

Nationalism functions in five ways. First, there is the matter of *identification*, whereby individuals consider themselves, especially since the advent of industrialization and its processes of urbanization and secularization, to be part of a nontraditional mass group, the "nation." Second, governments since the time of Napoleon have used nationalism as a means to *mobilize* military and economic power and to further their own legitimacy. Third, nationalism can function as a *centrifugal force* when it breaks up bigger countries (or empires) into smaller ones. This occurred in many European countries after World War I. It also took place in a massive way in the British Empire after World War II when India, Pakistan, Ghana, Nigeria, and other former colonies became independent. Then in the 1990s, it occurred again in the breakups of Czechoslovakia (into the separate Czech and Slovak republics) and Yugoslavia (into Slovenia, Croatia, Bosnia, Macedonia, and Serbia-Montenegro). Canada and Spain are two countries experiencing the centrifugal effects of separatist nationalism today: the French-speaking Quebecois in Canada and the Catalonians and Basques in Spain. The recent civil war in the Democratic Republic of Congo (formerly Zaire) is another important example of different national groups competing for power and gradually crippling the power of the central government.

Nationalism can also work in a fourth way, as a *centripetal force*, when it unites various people into new nation-states, such as occurred in the long and bloody unification struggles of the Germans and Italians in the nineteenth century, or in the Vietnam War of this century.

Fifth, nationalism can serve as a form of *resistance*, especially to colonial intruders. In Africa, the Middle East, and Asia, this has often been a kind of state-run, top-down nationalism that aims at organizing more meaningful resistance to actual or potential invaders. Sometimes this top-down nationalism is called "reform nationalism" (Breuilly 1993:9). In Turkey after World War I, Mustafa Kemal Ataturk launched a highly successful plan of economic and political modernization based on this kind of government-led reform nationalism. Cuba under Fidel Castro fits this definition as well. Another kind of resistance to colonialism takes the form of wars of independence (sometimes called national liberation struggles).

Important examples of this kind of national struggle include the Vietnam War and the Algerian war of independence (1954–1962).

■ NEGATIVE ASPECTS OF NATIONALISM

We have seen that nationalism can be an individual's sense of identity, a political allegiance, and a force for military and political change. Arising from these different levels of meaning are various negative effects of nationalism. Many of the conflicts in the world today originate in national disputes. A quick glance over the headlines shows warfare, ethnic conflicts, or genocide in Bosnia, Chechnya, Rwanda, Indonesia, Canada, South Africa, Macedonia, Cyprus, and Ireland. We can discuss these negative, conflict-producing effects of nationalism in terms of the following categories: imperialism, the glorification of the state, the creation of enemies, the overlap with religion, discrimination against minorities, and competing rights.

■ Imperialism

Self-confidence and group assertiveness, integral aspects of nationalism, can lead to arrogance or aggressiveness. Imperialism, which is the projection of a country's power beyond its borders to achieve the subjugation or exploitation of another country, is as old as history itself. But it takes on greater intensity when it meets with a sense of national unity and purpose. The "scramble for Africa" of the late nineteenth century, when many European states collaborated in literally carving up and occupying almost the whole continent, is a breathtaking example of arrogant imperialism imbued with a purported "civilizing mission" or "white man's burden," which justified the exploitation of other races. Carried to a much greater extreme, nationalism can end in genocide, as it did in the wildly homicidal policies of Adolf Hitler in the Third Reich, who sought to rid the world of Jews in order to make it "safe" for Germans.

■ Glorification of the State

Although many early nationalists, especially in the nineteenth century, believed that the nation-state was a vehicle of progress and liberty for all human beings, not all nationalist thought is connected with individual freedom. Indeed, nationalism often encourages antidemocratic practices. When a "people" or nation feels threatened by neighbors, or when it has a history of underdevelopment or division, political leaders can make the case for an authoritarian (antidemocratic) government. Sometimes, in the case of fascist governments, which are extremely authoritarian and stress anti-individualism, racial or national homogeneity, scapegoating, and militarism, the state or its leader comes to be regarded as the ultimate expression of the people's character and ambitions (Payne 1995; Weber 1964). Loyalty to governments like these is extremely dangerous because of their aggressive and intolerant policies.

■ Creation of Enemies

Another negative effect of nationalism can take place at the most basic level of self-identification. When people identify with one group, they often develop mistrustful or hostile feelings about people outside that group. Even neighboring states with a great deal in common can come to mistrust each other, as in the case of the recent fishing controversies between the United States and Canada. Similarly, countries with common political interests and similar economic systems—such as the United States and Japan—can develop deep misunderstandings based largely on national feeling.

■ Overlap with Religion

In some conflicts, such as those in Northern Ireland, the Middle East, and Bosnia, nationalism and religion cross paths in a very destructive way. In the current three-way struggle in Bosnia, between mostly Orthodox Christian Serbs, Roman Catholic Croats, and Bosnian Muslims, religion is a factor. Adding a religious dimension to nationalism can intensify divisive feelings; for instance, it can sanction killing—or dying—for a cause. Thus, it can make nationalists more fanatical and conflicts bloodier (Landres 1996).

■ Discrimination Against Minorities

Other difficulties arise when states or countries are actually constructed on national principles. Such principles hold that only members of a given national or ethnic group have the right to live in the new national state. Often a related principle tends to hold a lot of weight also: for example, only members of a particular ethnic group should enjoy the full benefits of citizenship. This creates a problem for minority groups, which are quite numerous in today's world. Major examples include the Hungarians in Romania and Slovakia, the Russians in the former Soviet republics (now independent states) of Estonia, Latvia, and Lithuania, and, until the creation of the Irish Republic, the Irish in the United Kingdom.

■ Competing Rights

Another negative aspect of nationalism lies in the competing rights and claims states make against one another. Three kinds are derived from or have a major impact on ethnic and minority questions. The first involves *historic rights*. These include claims by one national group to a certain piece of territory based on historical precedent. In other words, who was there first? This issue is hotly debated in Transylvania, a large portion of western Romania that has a substantial Hungarian population. In Bosnia, the competing parties of Serbs, Bosnian Muslims, and Croats have each tried to prove that they contributed more to the region's cultural heritage and, having set the tone for the region's culture, deserve to mold the region's political future now.

Next are *ethnic rights,* which address the question of who is currently in the majority in a given region. The contemporary setting—determined by population counts, polls, and votes of self-determination (such as in the Austrian province of Carinthia and in the Polish-German region of Silesia just after World War I)—is the decisive factor, not the complicated historical record of settlements, assimilation, immigration, and emigration.

The final claim can be referred to as *strategic rights*. Sometimes a state will claim a piece of territory simply because it needs that territory in order to be viable. This usually means the land is necessary for the country's defense or basic economic well-being. For example, after World War I, the new state of Czechoslovakia was given the Sudetenland region, even though it was heavily populated with Germans. This was done to provide the fledgling republic with a more mountainous, defensible border. Unfortunately, the Nazi dictator Adolf Hitler would later attack Czechoslovakia both diplomatically and militarily to "liberate" the Germans of that region, who he claimed were being denied their right to self-determination.

■ CHALLENGES TO NATIONALISM

We have seen that historically nationalism has been opposed by many forces. In addition to the conservative opposition in Europe, European imperialist powers in Africa—especially Great Britain, France, Belgium, and Portugal—resisted the growth of nationalism in their colonies. They did this despite being more or less nation-states themselves. This is because nationalism among colonized peoples presented a direct challenge to European domination and exploitation.

We now examine the four main challenges to the nation-state. One of these challenges is inherent in the ideal of nationalism itself. This is the problem of carrying the principle of self-determination through to its logical conclusions; if one national group deserves its own country and independence, then do not all groups deserve these things too? But countries are destabilized when every ethnic group within them agitates for its independence. And sometimes so-called microstates are created that are too small to be economically viable and that swell the membership of the United Nations and affect voting patterns there. For instance, the Pacific island country of Kiribati has about one-eighth as many people as the Canadian city of Toronto; likewise, the combined populations of thirty-eight microstates total only about a third of that of California (Rourke 1995: 201–202).

A related issue is *devolution*, or the decentralization of power in ethnically mixed countries. This usually does not result in the breakup of the country. The United Kingdom continues to experiment with this principle by giving more and more autonomy to its Welsh and Scottish regions. Belgium has also achieved a balance, based on this principle, between its Flemish and Walloon populations. Russia is faced with this issue today in many autonomous regions and districts.

Second is the issue of supranational groupings of various kinds. At the height of the era of decolonization, some Arab and African countries tried to establish political leagues that cooperated on a wide variety of issues. Today there are regional political and economic groupings on every continent. But in Europe, the blossoming of the European Union seems to herald an age of ever greater integration of nations. The United Nations, of course, while generally respecting the sovereignty of all countries, is the best example of a global grouping above the national level. Other contemporary examples include the North American Free Trade Agreement (NAFTA) and the Mercado Común de Sud-América (Mercosur); both are regional free trade groups in the Western Hemisphere.

Third, modern economic developments are also undermining the nation-state. The influence of multinational corporations, the obsession with free trade, and the appearance of a global, computer-driven, mass market economy are breaking down barriers between populations and eroding the sovereignty of smaller, less developed nations. This trend is analyzed in detail in Benjamin Barber's (1996) *Jihad vs. McWorld,* where a grim picture is painted of an increasingly standardized, shallow world culture dominated by a few, nearly all-powerful, marketing agents and producers of consumer goods.

Today the concept of national identity, and even to some degree the concept of nation-states, is in flux. The prevalence of computer-driven communication on the Internet and the World Wide Web affects society in many ways. From shopping to political discussions to dating networks, geography and distance are suddenly rendered virtually inconsequential by computers. The much-heralded "global village" of travel and communication has to some extent arrived, although its effects will likely never be as graciously positive and progressive as the gurus of technology predicted a few decades ago. Computer culture has developed rapidly along with the general economic shift in the world's most developed countries (such as Germany, Great Britain, the United States, Canada, and Japan) into a service-based economy (in contrast to economies based on the production of industrial goods). Many important changes in thought and attitudes go along with these technological and economic shifts. Service economies are oriented toward individual consumption, and the Internet means that individuals can have a maximum of "self-fulfillment" with a minimum of real contact with other citizens. This can reduce the sense of group loyalty so important in nationalism.

Finally, the world today is also witnessing a revival of conservative religious activity. This is most prominent in the Muslim world, but it is also present to some extent among other religions. Politically speaking, it is the new Islamism (sometimes called "Muslim fundamentalism") that most affects international politics, because it rejects capitalism and the decadence of Western culture as manifestations of a new imperialism. Since much, although by no means all, of the Muslim world consists of states that are ethnically and linguistically Arab, there is added potential for cooperation that transcends political boundaries.

The following two case studies illustrate national conflicts in various parts of the world today. They give us an idea of how national issues mix with other kinds of problems to create major crises.

■ CASE STUDY ONE: YUGOSLAVIA

The region of southeastern Europe known as the Balkans provides numerous intriguing case studies of nationalism at work. One of the characteristics of the region is the prevalence of ethnic or cultural nationalism rather than civic (or political) nationalism. Another is the highly diversified nature of its population. In many areas, numerous ethnic or national groups live close together; groups often intermingle and sometimes occupy the ancestral homelands of their neighbors. Two of the most mixed of these areas are Bosnia and Macedonia, both of which were part of the former Yugoslavia.

A third major characteristic of Balkan societies is a long history of foreign rule. Various empires, from the Ottoman and Hapsburg to the Russian and Soviet, have dominated the region, preventing the self-determination of its peoples. The two main peoples within the multinational state of Yugoslavia were the Serbs and Croats. It was their conflicting national aspirations—strengthened and made poisonous, many would say, by their current leaders—that provided the impetus for the breakup of the country in 1991–1992.

The term *Yugoslavia* means simply "land of the South Slavs." The country was created in 1918 as a kind of catch-all state for a number of small nationalities, including several that had been part of empires that collapsed in World War I. Thus, the term *Yugoslav* did not correspond to any genuine national or ethnic group; it was a matter of citizenship only, except for a small number of idealists or people who were part of mixed families created by marriages between members of different national groups.

During its existence, the country—first under the authoritarian rule of the Serbian royal family and then under the firm hand of the Communist military leader Josip Broz (known as Tito)—was divided into provinces or "republics" that reflected its chief national groups: Serbs, Croats, Bosnian Muslims, Slovenes, etc. There were also large and important minority groups, especially Albanians.

Rivalries among the various South Slavic national groups have been common, as they are among almost all neighboring peoples. But the frequently used journalistic phrases "ancient ethnic hatreds" and "long-smoldering ethnic feuds" are not accurate. While the Muslim-Christian rivalry in the Balkans had been a problem since the Middle Ages in Bosnia, the difficulties between Serbs and Croats became acute only during World War II.

After the Nazis and their allies carved up Yugoslavia in 1941, puppet states in both Croatia and Serbia emerged (see Figure 3.1). Both countries, but especially Croatia, sought to expand their territory and to homogenize their population at the expense of their neighbors and minorities. Further complicating the situation was the nature of Yugoslav resistance to the Nazis, which was led by the Communists under Tito but which included other rival political groups.

The post–World War II government sought to stabilize the country's national groups by one-party rule and by a decentralized administration. The three wars of succession that accompanied the breakup of the country after 1991—in Slovenia, Croatia, and Bosnia-Herzegovina—show that Tito's policies failed. There is also continued unrest among Albanians in Macedonia and in the Kosovo region of Serbia. In some ways, Tito may have made the national situation worse. Still, it is impossible to attribute the breakup of the country to any one cause. Nationalism played a part, as did economic problems, the ambitions of current leaders, and the failure of the Communists to allow or promote a pluralistic civil society that could have taught deeper loyalties to the central government and the Yugoslav ideal.

Figure 3.1 Yugoslavia and Its Successor States (Serbia, Montenegro, and Kosovo constitute what remains of Yugoslavia as of 1997)

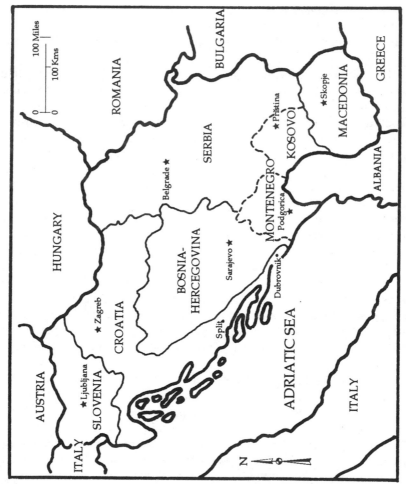

Source: Reprinted from Wayne C. McWilliams and Harry Piotrowski, *The World Since 1945: A History of International Relations*, 4th ed. (Boulder, CO: Lynne Rienner, 1997). © Copyright 1997 Lynne Rienner Publishers.

■ CASE STUDY TWO: THE ARAB-ISRAELI CONFLICT

The contemporary troubles between Arabs and Israelis form the main, but not only, crisis in the Middle East. This conflict centers on possession of the territory known as Palestine, which is important in both a religious and a historical sense to both Arabs and Jews. Since the diaspora (forcible dispersion) of Roman times, most Jews have lived outside this ancestral homeland. Arabs were the majority, though not the only, population group in Palestine during the intervening centuries. During World War I, the British, who were also cooperating with the Arabs, issued the Balfour Declaration in support of a Jewish state there. In 1948, when Palestine was partitioned and the state of Israel was created (see Figure 3.2), hundreds of thousands of Arabs were displaced. As a result, many Muslim states have, at least until recently, refused to recognize Israel's right to exist.

Wars between Israel and its Arab neighbors broke out in 1948, 1956, 1967, and 1973. Although Israel won all the wars, the situation remained volatile in large part because of the Cold War. The Soviet Union supported Arab states, while the United States aided the Israelis. In 1964, the Palestine Liberation Organization (PLO) was formed. Led by Yasir Arafat, the PLO has operated as a refugee organization, a government-in-exile, and a terrorist group; today it has been granted partial state power in several regions in and around Israel.

In addition to the territorial, religious, and linguistic differences between the Arabs and the Israelis, other factors have combined to make the nationalism of the conflicting parties more intense. Some Jews began to take an active interest in returning to their traditional homeland in the 1890s, due to increasing anti-Semitism in Europe. The movement known as Zionism proclaimed the Jewish people a nation with territorial claims to Palestine and not just a religion. In the 1940s, the terrors of the Holocaust gave added impetus to Jewish emigration.

On the Arab side, the issues involve more than just the fate of the Palestinian Arabs; many Arabs view Israel as an outpost of Western imperialism. Long dominated by foreign powers, the Arabs in this century have revealed "frustration over past and present weakness" (von Laue 1987: 350). Since World War II, the Arab countries have cooperated in various international bodies designed to foster unity and common purpose. These include the Arab League (since 1945), the Organization of Arab Petroleum Exporting Countries, or OAPEC (since 1960), and the short-lived United Arab Republic, a fusion of Egypt and Syria.

Like the Yugoslav case, nationalism has mixed with other historical and political issues such as population issues and Cold War rivalry. Given the complexity of this situation, it is no wonder that the main actors, even with the assistance of the international community, have not been able to solve the Palestinian issue.

Figure 3.2 The Expansion of Israel

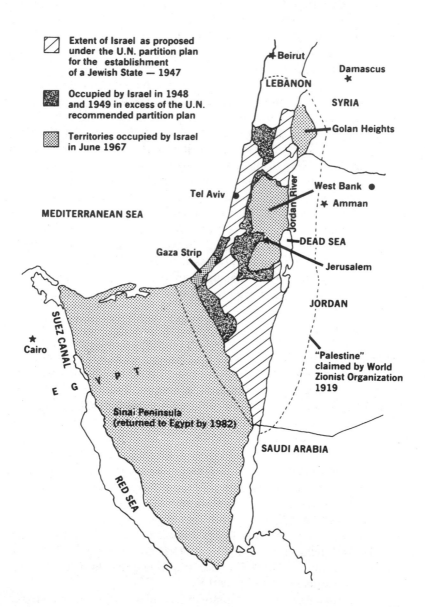

■ CONCLUSION

As we have seen, nationalism has been one of the most powerful forces in the world over the past 200 years. It operates in various ways on individuals, social groups, and governments. Nationalism can build states or destroy them; it can provide individuals or organizations with identity, motivation, and justification for their actions. Although it is true that nationalism provides important links between people and can give them a common purpose (as in patriotism), nationalism also causes conflicts and can be used to justify aggression and feelings of superiority. Thus, it has a very mixed reputation in terms of its positive and negative effects. While most people regard nationalism as an inevitable companion to social modernization, there are some signs that the usefulness or relevance of nationalism may be on the wane.

■ QUESTIONS

1. Which of the functions of nationalism do you think is the most important, and why?

2. What are some of the negative aspects of nationalism? Which do you think is the most dangerous or immoral?

3. Why do some countries have civic nationalism and others ethnic nationalism?

4. Which of the two types of nationalism is most prevalent in the United States?

5. What parts of the former Yugoslavia are most likely to break out in violence in the future?

■ SUGGESTED READINGS

Breuilly, John (1993) *Nationalism and the State*. Chicago: University of Chicago Press.

Brown, Michael E., Owen R. Coté, Sean M. Lynn-Jones, and Steven E. Miller, eds. (1997) *Nationalism and Ethnic Conflict: An International Security Reader*. Cambridge: MIT Press.

Eley, Geoff, and Ronald Grigor Suny, eds. (1996) *Becoming National: A Reader*. New York: Oxford University Press.

Geenfeld, Liah (1992) *Nationalism: Five Roads to Modernity*. Cambridge: Harvard University Press.

Hutchinson, John, and Anthony D. Smith, eds. (1994) *Nationalism*. New York: Oxford University Press.

———— (1996) *Ethnicity*. New York: Oxford University Press.

Ignatieff, Michael (1993) *Blood and Belonging: Journeys into the New National-ism*. New York: Noonday.

Liebich, André, Daniel Warner, and Jasna Dragovic, eds. (1995) *Citizenship East and West*. London: Kegan Paul International.

Sugar, Peter F., and Ivo John Lederer, eds. (1994) *Nationalism in Eastern Europe*. Seattle: University of Washington Press.

Teich, Mikulaš, and Roy Porter, eds. (1993) *The National Question in Europe in Historical Context*. Cambridge: Cambridge University Press.

Tilly, Charles, ed. (1975) *The Formation of National States in Western Europe*. Princeton: Princeton University Press.

4

Human Rights

D. Neil Snarr

On December 10, 1948, the United Nations (UN) unanimously approved the Universal Declaration of Human Rights. To date the UN has approved some 200 documents relating to human rights (such agreements are often referred to as "conventions"). The attention human rights have received since the signing of this document is tremendous. As one member of the U.S. House of Representatives and student of human rights has put it, "The defense of internationally recognized human rights has become the most universally accepted moral standard in the world today. Across the ideological spectrum, from the far left to the far right, there is agreement that the one unifying spiritual ideal in modern society is the enhancement and enforcement of human rights" (Drinan 1987: vii). This chapter looks at these rights—the controversies that surround them, the efforts to support them, and the many forces that inhibit their realization.

■ HUMAN RIGHTS AND THEIR ORIGIN

The idea of human rights certainly precedes the twentieth century, but the topic has received much greater attention since the founding of the UN in 1946 and the events that immediately preceded it. During the previous century, many issues such as slavery and women's rights became major concerns of citizens groups in Europe and North America. These movements often produced significant results, including the abolition of slavery and the realization of many rights for women. The events that immediately preceded the establishment of the UN and the approval of the Universal Declaration of Human Rights, however, were World War II and the genocide of Jews and other groups in Europe. Because of these indescribable events, the world community founded the UN in the hopes of avoiding such wars and violations of human rights in the future.

What are human rights? One UN document refers to them as "inalienable and inviolable rights of all members of the human family" (UN 1988: 4). According to one scholar,

> The very term human rights indicates both their nature and their sources: they are the rights that one has simply because one is human. They are held by all human beings, irrespective of any rights or duties one may (or may not) have as citizens, members of families, workers, or parts of any public or private organization or association. In the language of the 1948 declaration, they are universal rights. (Donnelly 1993: 19)

How do human rights come into existence? Generally, discussions of human rights have started in the UN General Assembly, where they are debated, given public scrutiny, and voted on. Passage of human rights conventions in the General Assembly is the easy part; it only takes a majority vote. After the General Assembly approves these conventions, they are opened for signatures by member states; and after a designated number of countries sign them, they are said to "come into effect." It often takes many years for this to happen.

After coming into effect, the agreeing countries are expected to pass laws, if they do not have such laws, that will ensure the observance and enforcement of the conventions. Eventually, it is hoped, all countries will approve such human rights laws and they will become international law. Some countries, however, have signed such conventions and made little progress toward institutionalizing them. Thus, at every juncture of this process is the possibility that countries will declare their commitment to these agreements and not enforce them. This graphically demonstrates the dilemma of human rights recognition and enforcement in the world community.

Why do countries sign these conventions and not enforce them? Simply because all countries of the world want to appear to other countries and persons to treat their citizens justly. As we will see later, this interest in being perceived as just and humane to one's citizens is an essential factor in the process of encouraging the observation of human rights.

Most Western countries are strong supporters of human rights and generally sign the conventions and establish mechanisms whereby they will be observed. The United States is an exception. In many cases, the United States either refuses to sign the conventions or signs them after attaching many conditions. This raises serious questions about the U.S. government's commitment to universal human rights.

It should be noted that the idea that international law be used to protect the rights of individuals is very recent. International law has been a body of law to regulate the relationships between countries. Only since World War II and the founding of the UN has international law been understood to also give protection to individual rights (Drinan 1987).

What Rights Have Been Identified as Human Rights?

One way to understand the many rights that have been approved by the UN and member states is to divide them into three "generations" or classes. These three generations have different origins and represent different views of human rights.

The first generation of rights is said to be *civil and political* and are contained in Articles 2 through 22 of the Universal Declaration of Human Rights. They focus on the rights of the individual and to some extent emphasize the responsibility of countries to refrain from unjustly interfering in the lives of their citizens. Examples of these rights are freedom from discrimination based on one's status, such as race or gender; the right to life, liberty, and security; freedom from slavery and torture; equal treatment by the law, including freedom from arbitrary arrest, detention, and exile; and the assumption of innocence until proven guilty. Originating in seventeenth- and eighteenth-century Western ideas, these rights found expression in the revolutions of France, Britain, and the United States. In the United States, we often view these as civil rights and equate them with "human rights." Internationally this generation of rights has received the greatest emphasis.

The second generation of human rights is referred to as *social and economic* rights. Contained in Articles 23 to 27 of the Universal Declaration, they have grown out of the Western socialist tradition. To some degree they have developed in response to what some consider to be the excessive individualism of the first generation of rights and the impact of Western capitalism and imperialism. Although they are elaborated in other conventions, in the Universal Declaration, they include the right to social security and work with fair remuneration; the freedom to join unions; the right to rest, leisure, and an adequate standard of living (including food, clothing, housing, and medical care) and an education; the right to the cultural life of one's community; and the right to the moral and material interests resulting from any scientific, literary, or artistic production of which one is author. More than the first generation of rights these necessitate a proactive government acting on behalf of its citizens. They establish an acceptable standard of living for all—that is, a base level of equality.

The third generation of rights is referred to as *solidarity* rights since they require the cooperation of all countries. Article 28 of the Universal Declaration states that "everyone is entitled to a social and international order in which the rights and freedoms set forth in this Declaration can be fully realized." These rights do not have the status of other rights and are in the process of being formulated. Burns Weston says the following about them:

> Three of these reflect the emergence of Third World nationalism and its demand for a global redistribution of power, wealth, and other important values: the right to political, economic, social, and cultural self-determination; the right to economic and social development; and the right to participate in and benefit from "the common heritage of mankind" (shared earth-space resources; scientific, technical, and other information and progress; and cultural traditions, sites, and monuments). The other three third-generation rights—the right to peace, the right to a healthy and balanced environment, and the right to humanitarian disaster relief— suggest the impotence or inefficiency of the nation-state in certain critical respects. (Weston 1992: 19–20)

These human rights grow out of the plight of the poorest two-thirds of the world, most of which were colonies of the Western countries and which generally remain poor.

■ ENFORCING HUMAN RIGHTS

For the most part, the enforcement of human rights is left up to individual countries. The United Nations passes human rights conventions and expects them to be followed; but the UN also supports the principles of self-determination and nonintervention for its member states (the principles of state sovereignty). Out of this contradictory situation has slowly emerged a worldwide discussion of human rights and some movement in the direction of serious human rights support.

■ The UN and Human Rights Implementation

The UN operates at four levels in supporting human rights (Farer 1992). First, it formulates and defines international standards by approving conventions and making declarations. This was initiated in 1948 with the approval of the Universal Declaration and later with the approval of two more major agreements that expand and clarify that document: the International Covenant of Political and Civil Rights, which is closely related to the first generation of rights; and the International Covenant of Social, Economic, and Cultural Rights, which is similar to the second generation of rights. (Both came into force in 1976, but many countries still have not signed them.) These three documents are referred to as the Universal Bill of Rights. Beyond these three major documents are the nearly 200 conventions that deal with more specific issues such as genocide, torture, children, refugees, women, and minorities. All these conventions further clarify the earlier documents and spell out the responsibilities of countries that have signed them.

Second, the UN advances human rights by promoting knowledge and providing public support. This is accomplished through the press, training sessions, scholarships, expert committees, worldwide and regional conferences, and special research. Recent regional and worldwide conferences include those on the environment, development, women, and population. Although these topics might not appear to be related to human rights, on closer examination the concern for human rights permeates the structure and content of these conferences and the UN itself (see Farer 1992: 232–233).

At the third level, the UN supports human rights by protecting or implementing them. Although the task of directly enforcing human rights is primarily left to the states themselves, the UN does become involved in various means of implementation. States that are party to conventions are to file reports on their actions. The Human Rights Commission, established in 1946, has been the central UN organization dealing with human rights and has also employed working groups and special rapporteurs. These rapporteurs pursue special human rights issues, such as disappearances of large numbers of persons; and blatant policies of unequal treatment, such as apartheid in South Africa prior to the early 1990s.

Finally, the UN has taken additional steps at enforcement that some consider to be "structural and economic aspects of human rights issues." This refers to support for the third generation of rights, such as economic development for the poorer countries of the world. This has taken a great deal of UN resources but does not receive the attention that more dramatic actions do (Farer 1992: 235). To these four categories of human rights support could be added the more recent and overt efforts in the Persian Gulf, Somalia, Rwanda, and the former Yugoslavia. These include boycotts of aggressor states (Serbia), military action against Iraq and the former Yugoslavia, military support for the delivery of humanitarian aid (Somalia), and the protection of refugees (Rwanda). These actions may or may not set a precedent for future UN actions, but they have elicited a great deal of discussion in the world community.

Thus, with reference to human rights, the UN serves primarily as an oversight organ and a forum where these issues are brought to public attention. In terms of action, the UN generally responds slowly and with uncertainty, subject to the interests of powerful governments. "The perspectives and capabilities of the United Nations are generally long term and indirect, so that direct, effective human rights protection remains an elusive goal" (Claude and Weston 1992: 220). Still, the UN has been somewhat successful in embarrassing violators into correcting their ways, and certainly the UN has become the setting for human rights dialogue.

It must not be overlooked, however, that there may be some cracks developing in the ideology of state sovereignty that will result in serious intervention in a state's internal affairs when massive human rights are violated. At the close of the Gulf War in April 1991, the Security Council passed Resolution 688, which was understood to permit the establishment of temporary havens for refugees inside Iraq. This in fact happened, and without the permission of Iraq. The rationale was that the violent treatment of Kurds (a large ethnic group living in Iraq) by the government of Iraq threatened international peace and security. A more recent example was the establishment of tribunals (or courts) to try persons responsible for crimes against humanity in the former Yugoslavia and Burundi. This is discussed later in this chapter.

▨ Human Rights Implementation
Outside the United Nations

The UN is clearly evolving with reference to issues related to human rights. This evolution must be seen in the total environmental context in which it exists. For example, the Cold War, which stretched from soon after World War II to 1990, created an atmosphere in which human rights concerns were clearly secondary considerations. Competition between the West led by the United States and the East led by the Soviet Union placed perceptions of national security above violations of human rights. Since the end of the Cold War, however, the hopes for greater attention and commitment to human rights has not been realized. Added to this are the many attacks against the UN in the U.S. Senate. The United States as well as a few other countries have refused to pay assessments (dues) they have paid in the past (the United States owes over a billion dollars). The United States was also recently successful in blocking the reappointment of UN Secretary-General Boutros Boutros-Ghali and demanded the restructuring of the international body. Although some of this is possibly justified, the United States is often perceived as a bully who will cooperate only when its demands are met.

Although the UN is legitimately seen as the primary forum for human rights discussions and states as the ultimate enforcers, it is expected that other domains of support will emerge. Two already have: regional human rights organizations, and private nongovernmental organizations (NGOs).

Three regions of the world now have human rights structures, but they are at very different levels of development. The most advanced and effective is the one in Europe that operates under the Convention for the Protection of Human Rights and Fundamental Freedoms, which was established in 1950 and functions under the European Commission of Human Rights. The commission receives complaints from approximately 4,000 individuals per year. After a process of analysis, this large number has been reduced to some forty, which the commission "pursues vigorously, and a majority end with a decision against the state" (Donnelly 1993: 82). Thus, "individual human rights" are realized, even though they are opposed by the laws of sovereign states.

A similar, but much less successful structure exists in the Americas. It includes the seven-member Inter-American Commission of Human Rights and the Inter-American Court of Human Rights. The commitment to human rights through these structures has been limited. "Its decisions . . . have usually been ignored, in sharp contrast to those of its European counterpart" (Donnelly 1993: 86). Countries of the Americas have resisted these institutions, and the principle of state sovereignty has predominated. The commission and court are left with the power of publicity and moral influence, which have been quite limited.

Finally, in the 1980s, African states approved the African Charter on Human and Peoples' Rights. It is a very interesting document in that, unlike other regional documents, it includes the rights of "peoples," or what is called third-generation or solidarity rights. Article 19 states, "All peoples shall be equal; they shall enjoy the same respect and shall have the same rights. Nothing shall justify the domination of a people by another." The document does not provide for a human rights court and emphasizes mediation, conciliation, and consensus.

Probably the most hopeful development in recent decades for defending human rights has been the activities of NGOs. Since the overwhelming majority of countries have expressed limited commitment to human rights, NGOs have come to serve the purpose of monitoring human rights violations throughout the world. Thousands of these organizations exist, but again they are more active and successful in areas where general support for human rights is evident. In several countries where human rights have made little headway, the very presence of NGOs is often unsupported, and sometimes their operations are prohibited. In many parts of the world, it is dangerous to work for human rights groups. For instance, a 1988 human rights publication reported that in 750 cases, human rights monitors were persecuted by governments and other armed groups (Human Rights Watch 1988).

NGOs take many forms and operate in many different ways. By operating outside of government they are able to monitor the actions of governments and bring pressure on governmental policies. Laurie Wiseberg lists nine areas in which NGOs provide services: (1) information gathering, evaluation, and dissemination; (2) advocacy to stop abuses and secure redress; (3) legal aid, scientific expertise, and humanitarian assistance; (4)

national and international lobbying; (5) legislation to incorporate or develop human rights standards; (6) education, conscientization (raising the consciousness of citizens), or empowerment; (7) solidarity building; (8) delivery of services; and, (9) access to the political system (Wiseberg 1992: 73–77).

Amnesty International, International League for Human Rights, Cultural Survival, International Commission of Jurists, International Committee of the Red Cross, Physicians for Human Rights, and many other groups provide services. Although they are somewhat limited in what they can do, they represent one of the most promising avenues for human rights support. The UN has made provisions for these organizations to have an official representation at the UN and at UN-sponsored conferences.

■ CURRENT ISSUES RELATED TO HUMAN RIGHTS

The emergence of human rights as a new issue on the world's political agenda in recent decades is clearly not without controversy. At every step since the signing of the Universal Declaration of Human Rights, there have been controversies, delays, and denouncements. There is no reason to believe that this will stop. In the sections that follow, we address a few of these controversies and the views that surround them.

■ State Sovereignty, Cultural Relativity, and Female Genital Mutilation

The UN Charter guarantees state sovereignty or self-determination and nonintervention; it also proposes that all individuals, regardless of their citizenship and status, have human rights. Countries jealously guard their sovereignty and at the same time profess that they grant their citizens human rights, which often they do not. States sometimes explain or rationalize their human rights violations by referring to the standards and values of their particular culture and traditions. This is known as *cultural relativity.*

Cultural relativity is viewed in a variety of ways (Donnelly 1993). "Radical relativists" see culture as virtually the sole source of values and would accept infanticide and female genital mutilation as cultural practices that should not be the concern of outside groups. At the other end of the spectrum are the "universalists," who view all human rights as universal. As the debate over what rights are universal proceeds, most decisions will fall somewhere between these two extremes. At the same time, however, there is general agreement that such acts as genocide (the killing of a people such as Jews, Gypsies, Hutus, or Tutsis), torture, and summary executions are violations of human rights.

Female genital mutilation (FGM) is a highly controversial human rights issue. This procedure

> usually involves the complete removal of the clitoris, and often the removal of some of the inner and outer labia. In its most extreme form—infibulation—almost all the external genitalia are cut away, the remaining flesh from the outer labia is sewn together, or infibulated, and the girl's legs are bound from ankle to waist for several weeks while scar tissue closes up the vagina almost completely. A small hole, typically about the diameter of a pencil, is left for urination and menstruation. (McCarthy 1996: 32)

This procedure affects some 100 million women in Muslim Africa and for different reasons is defended as a cultural tradition. Reasons given are that it makes girls "marriageable" (because it ensures their virginity) and also diminishes the sex drives of women. FGM is viewed by its proponents as a cultural practice that is a personal matter that should not be considered a human rights violation. Others consider this procedure a violation of a woman's human rights that should be eliminated.

The debate as to whether this is a matter of cultural discretion or a violation of human rights has taken place for many years, but in 1993, the World Health Organization (WHO), an agency of the UN, voted to take a stand against female circumcision. "A resolution approved unanimously by its 185-nation annual assembly asked WHO Director-General Hiroshi Nakajima to prepare a report that will call for tougher action on female circumcision and fasting during pregnancy" (*Columbus Dispatch* 1993). This has not put an end to the practices, but it is a small step taken by a UN agency to initiate what may eventually be recognized as a human right; states, in turn, will be challenged to face this reality. By making FGM an issue in the world press and among women's groups throughout the world, the practice has come into question. This is a very good example of how human rights issues reach the public, receive international attention, become a topic of debate, and possibly become codified in international law. This has already happened in several Western countries. It is clearly a very slow process.

If it were not for the international discussion of FGM, it is doubtful that Fauziya Kasinga, a native of Togo, would have received asylum in the United States after a year in a U.S. prison. After coming to the United States and admitting to immigration officials that she did not have a valid passport, it took a year of intense legal effort before the U.S.

> Board of Immigration Appeals granted political asylum to Fauziya Kasinga, recognizing FGM as a form of persecution against women. The ruling sets a binding precedent for all U.S. immigration judges. It also leaves open the possibility that women who have already undergone FGM may seek asylum in the future. (Burstyn 1995: 16)

Source: Jim Borgman, *Cincinnati Enquirer,* 1996. Reprinted with special permission of King Features Syndicate. © 1996 by King Features Syndicate.

■ Human Rights: A Form of Western Imperialism?

Does the controversy surrounding the imposition of Western human rights values on a non-Western tradition constitute a form of Western human rights imperialism? There is another region of the world where the cry of Western imperialism is heard with greater force. Several East Asian countries—China, Indonesia, and Singapore among them—view human rights as a product of Western civilization and not fully applicable to their societies. Although they do not refer to what are considered the more serious violations of human rights—such as murder, slavery, torture, and genocide—as controversial, they do take issue with first-generation rights such as freedom of the press, speech, association, and expression. They argue that they had little input into the Universal Declaration of Human Rights and that it expresses values they do not necessarily support (Bell 1996).

One argument poorer Asian countries give to support their position is that economic development necessitates at least the temporary suspension of some rights. They say that suspending certain rights will result in a greater good for more persons in the long run, when economic growth is realized. Western supporters of human rights respond that there is little evidence that human rights inhibits economic growth. Some argue that quite the opposite is true—that social and political rights may help ensure such economic growth.

East Asian countries also argue that Western human rights advocates overlook the negative consequences that emphasizing individual freedom brings. They point to the many social problems that the West, especially the United States, is experiencing. Such problems as drug abuse, crime, declining commitment to the family, homelessness, racism, and general alienation feed this skepticism (Faison 1997). They feel that the focus on first-generation rights as opposed to second-generation rights is at least partially responsible for Western moral decline. Further, they argue that the United States' unjustifiable involvement in the Vietnam War and its close relationships with and support for many Asian governments that have massively violated human rights are more reasons to be skeptical of Western-based human rights or at least their sincerity (Faison 1997).

■ China, the United States, Human Rights, and Trade

Countries may support human rights principles in their international dealings by passing laws that require that human rights records be considered before aid is given to countries. Canada, the Netherlands, and Norway follow this practice, as does the United States in many ways. A respected human rights scholar says the following about the growing importance of human rights issues in international dialogue:

> Human rights has become broadly important in contemporary international relations. In historical perspective, the changes of the last fifty years constitute an incremental revolution in the nature of international relations. Within the [extant] and resilient modified nation-state system, the broadly defined rights of persons matters as never before. Concern for human rights is intertwined with concern for state security, economic health, and a sound environment. Human rights has arrived as one of the major subjects of world affairs. (Forsythe 1991: 191)

This is certainly true with reference to economic relations between the United States and China. Since the United States is the largest consumer of goods in the world, all countries want access to its markets. For China this opening can come through most-favored-nation status and membership in the World Trade Organization. By granting most-favored-nation status to China, exports from China to the United States receive preferential tariff rates.

This discussion, much of which has gone on for years, is permeated with human rights language. In 1989, the Chinese government broke up a student prodemocracy movement in Tiananmen Square with ruthless disregard for life and was broadly condemned by the West. This was added to the charges that in China, arbitrary arrests are common and that the Chinese government disregards several other first-generation rights. These concerns have reached several of the international groups that consider human rights violations.

For six of the past seven years, the United States has cosponsored a resolution at the UN Human Rights Commission in Geneva to call for an investigation of China's human rights record. On all occasions China has vetoed the resolution. In 1996, however, the United States established some conditions under which the resolution would not be reintroduced. In turn, China was asked to accept International Red Cross–supervised prison visits, release a list of certain political prisoners, sign two UN human rights conventions and have them ratified, and resume a dialogue with the United States on human rights (Tyler 1997).

Recently the *New York Times* reported that China had reached an agreement with the Red Cross to reopen negotiations on access to prisons, something they have been unwilling to do since the establishment of the Communist regime nearly fifty years ago. The newspaper also reported that a senior Chinese prison official met with journalists and said that such visits by Red Cross officials are out of the question. The official went on to deny that there were any political prisons in China and added that China would not be subject to such regulation by a foreign entity such as the Red Cross, because such action is a violation of a state's sovereignty (Tyler 1997).

However, at the conclusion of that article, the author states, "Still, human rights advocates say, the diplomatic pressure has led to the release of hundreds of prisoners and has brought China closer to incorporating human rights concepts into its laws" (Tyler 1997). After recounting the legal changes that have taken place since Tiananmen, a researcher at Harvard Law School and human rights consultant said

> these new laws are far from perfect, and serious human rights abuses continue despite them. "But they are a real step toward rule of law in China—as opposed to arbitrary Communist Party rule—and they are being put in place because people inside China have an interest in that happening. You would never know that from the debate here." (Friedman 1997)

Thus, according to these sources, China is making progress toward the institutionalization of human rights at the same time that they are saying that these concerns are internal matters and purely state sovereignty issues.

Genocide in the Former Yugoslavia and Rwanda

There have been several cases of massive killings, including genocide, since the founding of the UN, the approval of the Universal Declaration of Human Rights, and other human rights conventions. For various reasons, however, they have not resulted in a significant response from the international community, including the UN. To a great extent this was due to the Cold War, which seemed to turn every event into a Cold War issue and subordinated human rights concerns to other interests. Donnelly (1993) points out that this was particularly true from 1948 to 1973 and during much of the period from 1981 to 1988. During these times, the United States supported many governments that were responsible for massive human rights violations but justified them on the basis of fighting communism. The USSR also had a poor human rights record, especially with respect to first-generation rights.

With the fall of the USSR and the end of the Cold War, there has been renewed hope for human rights, but this has also fallen short for various reasons. Two very serious cases of genocide have taken place during the 1990s, and the world community has been slow to respond. There has, however, been some movement, and this could prove to be significant. In early 1993 and again in late 1994, two international tribunals were established by the UN Security Council to deal with crimes against humanity in the former Yugoslavia and in Rwanda in central Africa. In the former Yugoslavia, some 200,000 people were killed in what was called "ethnic cleansing," and in Rwanda, approximately 800,000 were massacred in what appeared to be tribal violence.

The establishment of these tribunals is promising because it clearly illustrates that the international community is listening, even though it is acting slowly. (Massive killings that have occurred in Angola, Haiti, Liberia, El Salvador, Guatemala, Chechnya, Ethiopia, Somalia, and especially Cambodia in recent years received much less attention.) The establishment of the tribunals sets the cases of Rwanda and the former Yugoslavia off from the others, especially since they were established before the end of the conflicts. This constitutes a form of early intervention. Finally, and these are certainly not the only reasons to feel some hope, the establishment of these tribunals was based on international law, which supersedes state sovereignty. David Scheffer argues that since the UN Security Council established these two tribunals under the authority of Chapter VII of the UN Charter, it established an important precedent. International humanitarian law can subsequently override "domestic jurisdiction" (state sovereignty in such matters is challenged). If these tribunals succeed, they can become the tested and approved way for opening the door to judicial intervention, with the UN as the key actor. "Like human rights law in general international humanitarian law can reach deep into societies—into detention camps and remote villages where systematic torture or crimes against humanity are the sordid business of the day" (Scheffer 1996: 3).

Thus, if successful, these tribunals may serve as a message to regimes that commit massive rights violations that state sovereignty is not a facade behind which they can hide. On the other hand, both these tribunals have been justifiably criticized for their slow pace, which is due in part to inadequate funding and personnel. Clearly their success will depend on greater support of the world community, especially the United States. The Clinton administration has supported the tribunals and has been considering the establishment of a permanent UN human rights court.

■ Native Peoples and Human Rights

There are hundreds of millions of indigenous peoples throughout the world (for example, Navajos and Apaches in the United States, Inuits in Canada, Yanomamo in Brazil and Venezuela, and Maya in Mexico and Guatemala) whose well-being has been threatened for several centuries by Western expansion, population growth, state sovereignty, economic growth, resource exploitation, and a host of other forces. These peoples have found themselves subject to governments they do not recognize and forced into dilemmas not of their own choosing. What rights do *groups* of people have? Is there such a thing as group rights? Is there such a thing as "collective human rights"? This is another issue that challenges the principle of state sovereignty, since most indigenous peoples are subjects to modern states. "The focus of collective human rights . . . is on the rights of social groups, and proponents seek to create a normative framework independent of nation-states to enhance and project these rights" (Felice 1996: 18).

Many modern conflicts and wars are related to the rights of indigenous and ethnic groups that live within the boundaries of countries but experience little or no self-determination. Even before the founding of the UN, there were serious efforts to address the rights of such peoples, and the effort continues—much of it within the UN structure. For several years there has been work on a Draft Declaration on the Rights of Indigenous Peoples. This document speaks of self-determination within the present state system. For example, Article 27 states:

> Indigenous peoples have the right to the restitution of the lands, territories and resources which they have traditionally owned or otherwise occupied or used, and which have been confiscated, occupied, used or damaged without their free and informed consent. Where this is not possible, they have the right to just and fair compensation. Unless otherwise freely agreed upon by the peoples concerned, compensation shall take the form of lands, territories and within resources equal in quality, size and legal status.

Like most human rights issues, the document is controversial and finds strong support from some and absolute rejection by others. Many governments have already incorporated many of the document's principles into law and are moving toward indigenous self-rule. Among them are

Canada, Australia, New Zealand, Denmark, Norway, and Sweden. Other governments have not moved as far as self-rule but have adopted new laws to protect the rights of indigenous peoples.

■ CONCLUSION

The emergence of human rights as a global issue followed World War II and the massive killings that accompanied it. The forum for the discussion has been the UN, where the Universal Declaration of Human Rights was approved in 1948; some 200 related documents have also been approved.

Though support for human rights principles has been strong in the General Assembly, the actions of member states often have been very weak. This is because implementation of these principles is left to the countries themselves, who jealously guard their sovereignty. The struggle for human rights receives some assistance through regional human rights structures, but only Europe has reached a stage of seriously defending human rights.

NGOs have emerged as a serious link in monitoring human rights violations throughout much of the world. Though they have limited funds and power and are often opposed by and sometimes persecuted by governments, their impact is exceedingly important. By bringing violation to the public's attention and working on behalf of victims, they have helped make human rights an international issue.

Many governments are also strongly committed to human rights and shape their economic and foreign policies to reflect this. Few major international issues reach the public sphere that do not mention first-, second-, or third-generation human rights issues. Political repression, rights in the workplace, rights to a clean environment, rights to self-determination, and numerous other issues fill the news media and call citizens and organizations to action.

Progress toward a world in which human rights are recognized and enforced may be seen as a weak and nearly irrelevant issue, or as an emerging movement that will make the world a more livable place. The outcome is very much in the hands of individuals, organizations, and governments that daily make decisions that mold our future.

■ QUESTIONS

1. Which generation of human rights do you think is most important?

2. Why does the UN not enforce the human rights that the General Assembly has already approved?

3.   Should female genital mutilation be considered a cultural practice and
     not be condemned by human rights conventions?

4.   Which is most important: state sovereignty or universal human rights?

5.   Should indigenous groups have the right to self-determination even
     though they are located within the borders of a sovereign country?

■ SUGGESTED READINGS

Amnesty International Report (annual edition). London: Amnesty International
     Publications.
Claude, Richard Pierre, and Burns H. Weston, eds. (1992) *Human Rights in the
     World Community.* Philadelphia: University of Pennsylvania Press.
Donnelly, Jack (1993) *International Human Rights.* Boulder, CO: Westview Press.
Downing, Theodore E., and Gilbert Kushner, eds. (1988) *Human Rights and An-
     thropology.* Cambridge, MA: Cultural Survival.
Drinan, Robert F. (1987) *Cry of the Oppressed.* San Francisco: Harper & Row.
Felice, William F. (1996) *Taking Suffering Seriously.* Albany: State University of
     New York Press.
Forsythe, David P. (1991) *The Internationalization of Human Rights.* Lexington,
     MA: Lexington Books.
Gutman, Roy (1993) *A Witness to Genocide.* New York: Macmillan.
Human Rights Watch World Report (annual edition). New York: Human Rights
     Watch.
Staub, Ervin (1989) *The Roots of Evil: The Origins of Genocide and Other Group
     Violence.* Cambridge: Cambridge University Press.

5

Peacekeeping and Peacemaking

Carolyn M. Stephenson

Conflict issues now routinely cross international boundaries, blurring the distinction between domestic and international conflict. In response, methods of providing for international peace and security have evolved, especially in the post–Cold War world. Peacekeeping and peacemaking have taken new forms, and the role of multilateral institutions, especially the United Nations (UN), has become much more significant. What was initially optimism for a more peaceful world in the wake of the Cold War, however, has hardened into a recognition that we still face conflicts as intractable as the Cold War appeared to be from the late 1940s to the late 1980s.

In the late 1980s, we were full of hope for a new, more gentle world order that would provide for more peace and more security. There were hopes for a renewed UN, in new forms of mediation and other third-party conflict resolution, in a new relationship between the superpowers, in the signing of arms control and disarmament treaties, and in the increasing recognition of individual human rights and needs. Nonviolent revolutions had overturned authoritarian regimes in the Philippines in 1986 and in Eastern Europe in 1989. The UN had received the Nobel Peace Prize for its peacekeeping missions in 1988. We were beginning to develop solid international agreements to protect and restore our environment.

While we were less hopeful and more sober by the early 1990s, we had begun to take seriously both new dimensions of conflict and new approaches to peacemaking. But there was disagreement about what constituted the grounds for successful peacemaking. For some, success consisted of the breakdown of the Soviet Union and the renewed ability of the United Nations to function as originally intended as maker and keeper of the peace, under the leadership of the United States. Many of these saw U.S. military, economic, and political power as having been responsible for the end of the Cold War. For them, the restoration of the UN capability for enforcement action, under U.S. leadership, was central.

For others, what constituted success was the restoration of a different United Nations, a United Nations that would run by one-nation one-vote and counter the dominant influence of both superpowers, and that would function cooperatively to further individual and group rights, security, development, and the state of the global environment. For these people, U.S. leadership was not so central as that of decentralized political movements of individual human beings. For them, the changes had come about not so much because of U.S. military, economic, and political power as because of the committed organizing power of social movements all over the world.

The two approaches, and other variants of them, rely on different conceptions of security and peace and on different conceptions of power. And because they emphasize different methods of and approaches to peacemaking, it is important that we examine the conceptions of security, peace, and power that underlie these approaches. Otherwise, we risk shifting from notions of world government to arbitration, to nuclear deterrence, to rapid deployment forces, to UN peacekeeping, to nonviolent revolution, to arms control, to mediation, and on to humanitarian intervention, with no sense of why we have shifted to emphasize one approach over another, let alone of what the strengths and weaknesses of each are in particular situations.

■ CHANGES IN THE CONCEPT OF SECURITY

Where national security was once virtually the only way to talk about security, we have come to acknowledge the relationship between national security and both international and individual security. We have moved from reliance on a balance of power system, to collective security, to collective defense, and then to common security, with the present international security system some mixture of all of these.

The classical *balance of power* system, the primary system for maintaining security in nineteenth-century Europe, was retained well into the twentieth century. With a goal of ensuring that no nation-state became so strong as to be able to overpower others, rough equality was maintained by two camps of states in the system, with one or several states (usually Britain) changing alliances in order to maintain the balance. This system began to break down in the twentieth century, when it failed to avert war and maintain stability in the system.

Under *collective security*, which began with the League of Nations in 1919 and was strengthened with the establishment of the United Nations in 1945, states agreed on certain rough rules of international law, including national sovereignty and freedom from outside aggression; they also agreed that if any state violated these rules, all the others would band together against that state. Sanctions for violating the prohibition against international aggression could be either military or nonmilitary.

Collective defense, which was a step back in the direction of the balance of power system, became the dominant security system by the late 1940s. Under this system, each set of nation-states, West and East, gathered together in military alliance to defend against the other set. The formation of the North Atlantic Treaty Organization (NATO) in 1949 was followed by the formation of the Warsaw Pact in 1955. Each side bolstered its conventional military defenses with the nuclear umbrella of its respective superpower. While collective defense has not entirely ended, the withdrawal of Soviet forces from Eastern Europe in the late 1980s, and the end of the Warsaw Pact in 1990, left NATO with questions about its remaining purpose. Deterrence, including but not limited to nuclear deterrence, was the primary underlying power dynamic of collective defense.

In contrast, the concept of *common security* arose, primarily within the UN framework. There are two distinct aspects of common security, one of which arose in the context of North-South conflict, one in the East-West context. The Report of the Independent Commission on International Development Issues in 1980, better known as the Brandt Report, raised notions of economic security (Independent Commission on International Development Issues 1980). For the South, the failure of economic development was perceived as a much greater threat to security than nuclear war. The Independent Commission on Disarmament and Security Issues, or the Palme Commission, in 1982 made two more explicit linkages: first, there could be no victory in nuclear war—therefore we could only survive together; second, the costs of the military everywhere were contributing to economic insecurity—therefore the reduction of military costs could contribute to development (Independent Commission on Disarmament and Security Issues 1982).

The Conference on Security and Cooperation in Europe (CSCE), or the Helsinki Agreement, formed in 1975 by the states of both NATO and the Warsaw Pact and now known as the Organization for Security and Cooperation in Europe, is the best example of an actual common security regime. The Helsinki Agreement contains three "baskets": a security basket, which contains agreements on post–World War II borders in Europe; an economic basket, which opens up trade between East and West; and a human rights basket, which provides for certain human rights guarantees and procedures in both East and West. Since the time of CSCE, proposals have been made and other regional conferences held on security and cooperation; some governmental and some nongovernmental. These steps have been taken as confidence-building measures in regional conflict.

The Brundtland Commission report, *Our Common Future*, in 1987 added the concept of *environmental security* to that of common security, strengthening the idea that sustainable development required sustaining the environment that supported development and promoting the notion that military expenditure and war were harmful to the environment. Environmental security encompassed both the protection of the environment for its own sake and the protection of the environment for the sake of humankind. Common security now encompasses political-military aspects, economic aspects, and environmental aspects and seems to be recognizing their interdependence; but it is more explicit in acknowledging the interdependence of the security of states. Security today is thus conceived of in a much more comprehensive way, even when that security is still for the nation-state rather than the global society. It comprises not only negative security (the ability to defend against or shut off relationships one views as harmful), but also positive security (the ability to maintain relationships one views as essential to one's survival, such as access to food, oil, and credit). Such a reconceptualization of security to include both positive and negative security means that reliance on traditional approaches to security are less likely to be adequate. This is one of the reasons new approaches to peacemaking are increasingly being taken seriously.

■ CHANGES IN THE CONCEPT OF PEACE

The concept of peace has also broadened in much the same way as security has, expanding from the concept of "negative peace" alone—or peace as the absence of war or direct violence—to include "positive peace"—or peace as the absence of exploitation and the presence of social justice. While the earliest mention of positive and negative peace appears to be in the writings of Martin Luther King, Jr., the terms were expanded upon and more fully analyzed and operationalized by Johan Galtung (1969). The debate that ensued over which concept of peace was to be accepted has yet to be resolved. For some, the absence of direct violence seems more important; for others, the absence of exploitation is key. But until we are more in agreement about the kind of peace we are interested in making, there will continue to be major differences in the various approaches to peacemaking.

At one end of the spectrum, those who believe that peace is simply order or the absence of violence may argue that peace can or should be enforced with military action. At the other end of the spectrum, those who believe that peace must include justice may argue that peace cannot be enforced and can only be brought about by negotiations that take into account justice and the underlying needs of the parties. The international system has mechanisms that span the full range of these points of view.

For a long time, distinctions have been made in the UN between peacemaking, peacekeeping, and peace-building (Boutros-Ghali 1992).

Without getting into an argument over technical definitions, let it suffice to say that peace-building generally includes building the conditions of society so that there will be peace. In this area we might include such methods as human rights education, development and development aid, and reconciliation and the restoration of community following a violent conflict. Peacekeeping, in the broader sense, involves keeping parties from fighting or otherwise doing harm to each other. In the narrower sense, it has been used to describe the particular multinational operations employed to restore and maintain peace between hostile parties. Peacemaking is usually taken to mean helping bring hostile parties to agreement. Let us explore all three of these, with an emphasis on peacekeeping and peacemaking and with the recognition that they overlap somewhat.

■ PEACEMAKING AND THE UNITED NATIONS CHARTER

The UN Charter includes two primary ways of providing for peacemaking and for international peace and security. Chapter VI of the Charter focuses on Peaceful Settlement of Disputes, while Chapter VII relates to Action with Respect to Threats to the Peace, Breaches of the Peace, and Acts of Aggression. Article 33, the first article of Chapter VI, provides that

> the parties to any dispute, the continuance of which is likely to endanger
> the maintenance of international peace and security, shall, first of all,
> seek a solution by negotiation, enquiry, mediation, conciliation, arbitra-
> tion, judicial settlement, resort to regional agencies or arrangements, or
> other peaceful means of their own choice.

When parties are unable to negotiate their way through a dispute on their own, the Security Council may call upon them to settle their dispute by any of these means, and it, or other parts of the United Nations, may assist them by providing a third party who may help them do so. *Enquiry* and *fact-finding* are methods by which a third party attempts simply to find out the facts of the situation. In *mediation and conciliation*, a third party, sometimes in the form of a special representative of the Secretary-General, sometimes in the form of a commission, helps the conflicting parties communicate and come to agreement when they are unable to do so. Under *arbitration*, the third party makes a decision about the conflicting claims of the parties; sometimes the parties can choose whether to accept this decision, but under binding arbitration, the parties are bound to accept the decision of the arbitrator. Under *judicial settlement*, the Charter provides for submitting certain types of legal disputes between states to the International Court of Justice in The Hague, where the court will rule on the legitimacy of the case under international law.

Mediation has become increasingly important in resolving international disputes. The "quiet diplomacy" of the UN Secretary-General or his representatives has been used in a long series of crises in the Middle East, beginning in the late 1940s and continuing through to the Iran-Iraq War, as well as in other regional areas. Mediation has also been used by international regional organizations such as the Organization for African Unity, as well as by powerful states such as the United States—for example, when it mediated the peace between Israel and Egypt during the Camp David negotiations—and by less powerful states such as Algeria, which mediated the release of the U.S. hostages held by Iran. Mediation is also practiced in the form of what is called "second-track diplomacy," where individuals such as academic specialists in conflict resolution, or representatives of the International Committee of the Red Cross or of religious organizations such as the Mennonite Conciliation Service or the Friends World Committee for Consultation (Quakers), help to facilitate communication or to run workshops aimed at solving the problems underlying the conflict.

UN enforcement action constitutes another approach to peacemaking that has become more available to the international system since the end of the Cold War. Enforcement action is covered under the collective security provisions of Chapter VII of the UN Charter, especially the *nonmilitary sanctions* provided in Article 41 and the *military sanctions* provided in Article 42. While some consider that the Charter conditions for military enforcement have never been met, due to the failure to set up UN forces under a Military Staff Committee (Articles 43–47), most would agree that UN-authorized operations in Korea in 1950 and in Iraq/Kuwait in 1991 constitute the primary examples of UN military enforcement. Enforcement action is generally considered when a state has clearly violated the terms of the Charter and carried out international aggressive action. Military enforcement actions have tended to be led and staffed by one or several of the great powers. There are clear differences of opinion as to whether military enforcement action constitutes an approach to peacemaking or is better considered simply as war.

Nonmilitary sanctions are another approach to peacemaking. Article 41 of the UN Charter says that "the Security Council may decide what measures not involving the use of armed force are to be employed to give effect to its decisions," including "complete or partial interruption of economic relations and of rail, sea, air, postal, telegraphic, radio, and other means of communications, and the severance of diplomatic relations."

The old debate over whether sanctions are appropriate and effective was renewed after Iraq's invasion of Kuwait. Peace organizations generally supported sanctions as an alternative to war, before the war, but opposed sanctions as harmful to the Iraqi people after the war. If the purpose was to get Saddam Hussein out of Kuwait, there were early indications that sanctions might have worked. If the purpose, on the other hand, was to get Saddam Hussein out of Iraq, sanctions were not likely to be effective. The ambiguity between those two goals in messages from the United States may be one of the reasons sanctions did not succeed within the short time period they were given.

In the 1930s, sanctions were seen as a primary guarantor, within the system of collective security, for preventing wars. After the failure of the League of Nations sanctions against Italy in 1936 with respect to Ethiopia, sentiment turned against sanctions. Most writers today have concluded that sanctions are not especially useful on major foreign policy goals but may be useful on limited goals.

The definition of *sanctions* has changed over time. By the time of the League of Nations, sanctions meant actions taken by international bodies to enforce international law. Since then, the term has come to include unilateral acts and even the use of economic policies for ordinary diplomatic influence. Evaluation of the success of sanctions may be very different if one separates out the unilateral from the more consensually based actions of international organizations.

■ PEACEKEEPING

On October 10, 1988, UN Secretary-General Pérez de Cuéllar accepted the Nobel Peace Prize on behalf of the 10,500 members of peacekeeping forces. He paid tribute to the half million young men and women from fifty-eight countries who served in UN peacekeeping operations, and especially to the 733 who had lost their lives. Both he and Norway's prime minister, Gro Harlem Brundtland, expressed their concern for the financial status of peacekeeping, particularly the fact that major powers are in considerable arrears. At that time, the annual cost of peacekeeping activities was about $300 million. By 1994, annual UN peacekeeping expenditures had reached $3.6 billion. Peacekeeping had come of age but was endangered by the massive costs that accompany certain of its aspects.

There is no official UN definition of *peacekeeping*. However, the definition adopted by the International Peace Academy (IPA), a nongovernmental organization (NGO) closely related to the UN that has undertaken much of the training for UN peacekeeping forces, has been seen as close to official. Under that definition, peacekeeping is

> the prevention, containment, moderation and termination of hostilities between or within states, through the medium of a peaceful third party intervention organized and directed internationally, using multinational forces of soldiers, police, and civilians to restore and maintain peace. (IPA 1984: 22)

A wide range of interpretations are still possible under that definition. UN publications distinguish between two kinds of peacekeeping operations: observer missions and peacekeeping forces. Observers are not armed, while soldiers in peacekeeping forces have weapons, but generally are authorized to use them only in self-defense. By 1990 there had been eight peacekeeping forces and ten observer missions. By June 1996 there had been forty-one peacekeeping operations (see UNDPI 1990: 3; UNDPI 1996). Peacekeeping is not based on sending a fighting force to stop a violent conflict. Rather, the premise of peacekeeping is that inserting an impartial presence in the region will allow the parties to try to negotiate a peaceful settlement to the conflict. There are differences in emphasis between the military and civilian role in peacekeeping operations, and different proportions in each operation, to achieve the double objective of reducing the violence and helping to move toward peaceful settlement.

Peacekeeping is clearly different from enforcement action or action based on collective security. The separation of peacekeeping from enforcement is critical as both peacekeeping and enforcement are made increasingly possible by the condominium of the great powers in the aftermath of the Cold War. The fact that peacekeeping missions are deployed in countries with their consent, that they are unarmed or lightly armed and use force only in self-defense, and that they are composed largely of middle-level powers whose degree of neutrality in the conflict is likely to be perceived as higher than that of the great powers, are significant factors that may well be important to their success.

◼ The Evolution of UN Peacekeeping

The changing world situation in the late 1980s led not only to an increase in peacekeeping, but also to the first UN Security Council summit meeting. In the concluding statement of the summit on January 31, 1992, the heads of state invited the Secretary-General to prepare "an analysis and recommendations on ways of strengthening and making more efficient within the framework and provisions of the Charter the capacity of the United Nations for preventive diplomacy, for peacemaking and for peacekeeping" (Boutros-Ghali 1995: 117–118). Secretary-General Boutros-Ghali, in his resulting report, spoke of the increasing demands for peacekeeping:

> Thirteen peace-keeping operations were established between the years 1945 and 1987; 13 others since then. An estimated 528,000 military, police and civilian personnel had served under the flag of the United Nations until January 1992. . . . The costs of these operations have aggregated some $8.3 billion till 1992. . . . Peace-keeping operations approved at present are estimated to cost close to $3 billion in the current 12-month period. (Boutros-Ghali 1995: 57–58)

A mechanism not originally provided in the UN Charter had evolved over time to become the major item in the UN budget, with the exception of the specialized agencies. How did UN peacekeeping evolve?

As definitions of peacekeeping differ, so consequently do writers differ on when UN peacekeeping began. For some analysts who include observer missions (e.g., Wiseman 1983), UN peacekeeping began with the UN Special Committee on the Balkans (UNSCOB), created by the UN General Assembly (UNGA) in 1947 to investigate the Greek allegations of Albanian, Bulgarian, and Yugoslavian guerrilla infiltration (UNGA 1947). For others (e.g., Higgins 1996), including the UN itself, peacekeeping began with the UN Truce Supervision Organization (UNTSO) in Palestine, which emanated from a 1948 Security Council resolution but was to be under the authority of the Office of the Mediator created by the General Assembly (UNGA 1948; UNSC 1948). UNTSO is one of the UN peacekeeping observer missions that still exists. Those early operations were the predecessors of later peacekeeping missions. These small early missions gradually developed a separation between the political mediation side and the military operations but maintained communication between them.

The term *peacekeeping* was created to describe the first UN Emergency Force (UNEF), which was established in Egypt in 1956 to secure the cease-fire and to provide face-saving for the withdrawal of Israeli, British, and French troops in the Suez crisis. A deadlock in the Security Council threw the conflict into the General Assembly under the 1951 Uniting for Peace Resolution. (This resolution allowed the General Assembly to take up security matters if the Security Council was unable to do so because it was blocked by a veto by one of the major powers.) The idea of a United Nations emergency force was first presented to the General Assembly in the wee hours of the night of November 1–2, 1956, by Lester Pearson of Canada, and received the support of the United States. Then, under the authority of General Assembly resolutions, the Secretary-General brought together the representatives of Canada, Norway, Colombia, and India to produce a plan for the establishment of a United Nations command under the leadership of Major-General E. L. M. Burns, a Canadian who was UNTSO chief of staff, with personnel to be recruited from states that were not permanent members of the Security Council. Several days later the plan passed the General Assembly by a vote of 57 to 0, with Egypt announcing acceptance of the plan the next day (UNGA Resolutions 997, 998, 999, 1000, and 1001). Parts of the new force were on the ground within a week. Although the United States was excluded from sending troops, it did contribute roughly 40 percent of total UN authorized expenditures for UNEF.

UNEF I was to be the first of a series of larger UN military operations that eventually led to a financial and constitutional crisis that threatened the entire UN. Lester Pearson had spoken in the Canadian House of Commons debates of discussion as early as 1953 on replacing UNTSO with a police force that would have greater authority. The Suez crisis brought such a force into being, with UNEF I having a maximum force level of 6,173.

The debate over the degree of military force the UN would control, and the method of that control, had arisen first in 1945 in San Francisco. The debate moved toward increasing military authority for the UN with UNEF in 1956 and the Organisation des Nations Unies en Congo (ONUC) in 1960. It arose again at the close of the Cold War, when the common interests of the former superpowers and major powers led to renewed consideration of both large and small peacekeeping missions, as well as military forces for enforcement actions under collective security.

UNEF I was followed by an even larger force of 19,825 in the July 1960–June 1964 ONUC operation. Congolese independence from Belgium on June 29, 1960, was followed in rapid succession by a Congolese Army mutiny, the return of Belgian troops, the proclamation of the independence of Katanga province, and a Congolese request for UN military assistance. With a mandate considerably more vague than that of previous UN forces, the UN Security Council agreed on July 14, 1960, to provide military assistance, with the Secretary-General clarifying the mandate on July 18 to indicate that restoration of order was the main task, that the force was exclusively under UN command and would not become a party to internal conflict, that military units would be drawn largely from African states and not permanent members of the Security Council, and that they would use force only in self-defense. Over the course of the operation, many of these tenets were violated as the UN force was drawn into what became a civil war. In spite of Secretary-General Dag Hammarskjöld's mediation efforts and the creation of a Conciliation Commission, the central part of the Congo operation became the UN use of force.

Financing became a problem. While the Soviets originally supported the operation, they later became severe critics and refused to pay financial assessments associated with the operation. This, coupled with a French refusal to pay in the UNEF I force, led to an International Court of Justice advisory opinion and a major financial crisis for the UN. Where peacekeeping had arisen out of the inability of the Military Staff Committee to function and because of the Cold War use of the veto in the Security Council, it had now received its most severe challenge by what was in effect a financial veto, and not only across Cold War lines.

In the same period, however, the UN Peacekeeping Force in Cyprus (UNFICYP) was launched in March 1964, achieving a maximum size of 6,411 and continuing to the present. UNFICYP has generally been considered a success in spite of the Turkish invasion in 1974 and the failure to move toward any settlement. It does, however, raise the question of whether the stability provided by a peacekeeping force can forestall the pressures necessary to bring parties to agree to negotiate.

The combination of the financial and neutrality questions that arose from the Congo operation, and the withdrawal under fire of UNEF I in 1967 at the demand of Egypt, led to the first dormant period in UN peacekeeping. Also, the Secretary-General, Dag Hammarskjöld, who had been so significant in the development of both peacekeeping and "quiet diplomacy" and their effective linkage, had been killed in an airplane accident in the Congo in 1961, leaving a leadership gap in this area.

A brief resurgence for peacekeeping came after the October 1973 war in the Middle East with the creation of the second UNEF from 1973 to 1979, the UN Disengagement Observer Force (UNDOF) in 1974 at the contested Israeli-Syrian border, and the UN Interim Force in Lebanon (UNIFIL) in March 1978. UNEF II was superseded by the arrangements under the Camp David agreement negotiated by the United States between Israel and Egypt. It has been widely regarded as successful during its period of operation, but it may be even more significant in terms of the cri teria for successful peacekeeping that the Secretary-General laid down at the time of its creation. These will be discussed below. Of the other two operations, both continue. UNDOF is widely regarded as one of the most successful, because it has continued to reduce tensions in the Golan Heights area, and UNIFIL is viewed as one of the least successful, having been overrun by Israeli armies in June 1982. No new peacekeeping operations were launched in the decade between 1978 and 1988, but in 1988 a whole new era of peacekeeping arose for the UN.

■ PEACEMAKING AND PEACEKEEPING
AFTER THE COLD WAR

At the end of the Cold War, when the focus of conflict turned from the threat of nuclear war to ethnic conflict in the developing world and elsewhere, there was an enormous rise in peacekeeping operations. Figure 5.1 indicates the geographic spread of both completed and ongoing UN peacekeeping operations from 1948 to the end of 1996.

Table 5.1 (pp. 74–75) shows that the sixteen current operations vary greatly in start date, size, budget, and purpose, as is revealed to some degree in the title of each mission. While a few of the current operations are almost as old as the UN itself, most began in the 1990s.

Many described these conflicts as a new type, although in reality they were not significantly different from the hundreds of intrastate ethnic, religious, racial, tribal, and national conflicts that had been occurring with

Figure 5.1   United Nations Peacekeeping Operations

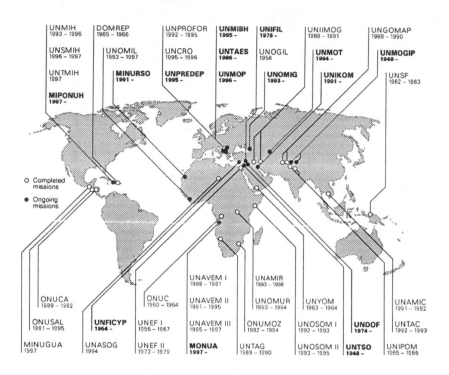

Source: United Nations Department of Public Information in consultation with the Department of Peacekeeping Operations and the Peacekeeping Financing Division (OPPBA), DPI/1634/Rev. 7, December 16, 1997. Used by permission of the United Nations.

Notes: Peacekeeping operations 1948–December 1, 1997 total 46.

Operations underway on December 1, 1997 total 15.

Personnel as of November 30, 1997: Military and civilian police personnel serving total 16,849 (includes 12,694 troops, 1,128 military observers, and 3,027 civilian police); countries contributing military and civilian police personnel total 71; fatalities among personnel assigned to 46 peacekeeping operations from 1948–November 30, 1997 total 1,530 (includes 1,407 military, 50 civilian police, 42 civilian international staff, and 31 local staff).

Financial aspects: Estimated total cost of operations from 1948–June 30, 1997 total about $17.3 billion; estimated cost of operations from July 1, 1997–June 30, 1998 total about $1.1 billion; outstanding contributions to peacekeeping on November 30, 1997 total about $1.6 billion.

regularity throughout the Cold War. The difference was that throughout the Cold War period the two sides had interpreted them as conflicts involving a competition between communism and capitalism, while now they were seen simply as intrastate conflicts.

The United Nations Security Council, because of the end of the Cold War and the pattern of Cold War vetoes there, also found that it could make decisions and take action it could not have taken before. This led not only to a tremendous rise in the number of peacekeeping operations, but also to a blending of traditional peacekeeping with enforcement in some cases and with elections monitoring, human rights monitoring, and education and other techniques in others. These new types of operations came to be called second-generation peacekeeping. Missions such as those in Cambodia, El Salvador, Haiti, and Mozambique are widely regarded as successes; they helped oversee elections and rebuild societies where there was agreement to do so. On the other hand, in places suffering a breakdown of society, rioting and looting, and a continuing struggle for power between factions—such as in Somalia and in the former Yugoslavia—action by the UN, other groups, the United States alone, or NATO has not been regarded as entirely successful. It has become increasingly clear that there must be a peace to keep before peacekeeping can be useful. Following the tremendous rise in peacekeeping in the early 1990s, the UN entered a period of reassessment and began to step back from its enormous investment and to draw conclusions on the lessons learned from peacekeeping operations. Among these was that it was essential to have the size and nature of the force fit the situation. The UN also began to realize that peace could not be enforced from the outside and that there was no substitute for political negotiations.

■ THE CONDITIONS OF SUCCESS

The chief conditions of success in peacekeeping, whether for restraining violence or for resolving conflict, are more likely to be political than military. A clear mandate adopted with the greatest degree of consensus possible and the consent of the parties to the conflict may be among the most important conditions. But there are other conditions that have developed out of the experience of UN peacekeeping.

The set of operating conditions that the UN Secretary-General attached to the deployment of the second United Nations Emergency Force in 1973 has come to be regarded as prerequisites for success. Deployment must be done only

1. with the full confidence and backing of the Security Council, and
2. with the full cooperation and consent of the host countries.

The force must

3. be under UN command through the Secretary-General,
4. have complete freedom of movement in the countries,
5. be international in composition, with national contingents acceptable to the parties in conflict,
6. act impartially,

Table 5.1   Current Peacekeeping Operations

UNTSO   Since June 1948
United Nations Truce Supervision
Organization
Strength[a]: 156
Fatalities[b]: 38
Budget estimate for 1997: $23.7 million

UNMOGIP   Since January 1949
United Nations Military Observer Group in
India and Pakistan
Strength: 44
Fatalities: 9
Budget estimate for 1997: $6.4 million

UNFICYP   Since March 1964
United Nations Peacekeeping Force in Cyprus
Strength: 1,267
Fatalities: 168
Budget appropriation July 1997–June 1998:
$48 million (gross)
UN assessment on Member States: $29
million; and voluntary contributions by
Cyprus ($15.3 million) and Greece ($3.7
million)

UNDOF   Since June 1974
United Nations Disengagement Observer
Force
Strength: 1,048
Fatalities: 39
Budget appropriation July 1997–June 1998:
$33.6 million (gross)

UNIFIL   Since March 1978
United Nations Interim Force in Lebanon
Strength on November 30, 1997: 4,475
Fatalities: 222
Budget appropriation July 1997–June 1998:
$125 million (gross)

UNIKOM   Since April 1991
United Nations Iraq-Kuwait Observation
Mission
Strength: 1,049
Fatalities: 10
Budget appropriation July 1997–June 1998:
$51.5 million (gross)
Two-thirds of the cost ($33 million) is paid
by Kuwait

MINURSO   Since April 1991
United Nations Mission for the Referendum
in Western Sahara
Strength: 242
Fatalities: 7
Budget appropriation July 1997–June 1998:
$30.2 million (gross)
Proposed additional requirements requested
due to resumption of identification process:
$17.9 million (gross)

UNOMIG   Since August 1993
United Nations Observer Mission in Georgia
Strength: 106
Fatalities: 3
Budget appropriation July 1997–June 1998:
$18.6 million (gross)
Indicated financial implications related to the
planned expansion of UNOMIG: about
$1.7 million

UNMOT   Since December 1994
United Nations Mission of Observers in
Tajikistan
Strength: 44
Fatalities: 1
Budget appropriation July 1997–June 1998:
$8.3 million (gross)
Indicated financial implications related to the
planned expansion of UNMOT: about
$13.7 million for 6 months

UNPREDEP Since March 1995
United Nations Preventive Deployment Force
Strength: 928
Fatalities: 4
Budget appropriation July 1997–June 1998:
$46.5 million (gross)

UNMIBH   Since December 1995
United Nations Mission in Bosnia and
Herzegovina
(Incorporates International Police Task Force,
IPTF)
Strength: 2,047
Fatalities: 10
Budget appropriation July 1997–June 1998:
$189.5 million (gross)

(continues)

Table 5.1 continued

UNTAES Since January 1996 United Nations Transitional Administration for Eastern Slovakia, Baranja, and Western Sirmium Strength: 11,226 Fatalities: 10 Budget appropriation July 1997–June 1998: $275.4 million (gross) UNMOP Since January 1996 United Nations Mission of Observers in Prevlaka Strength: 28 Cost included in UNMIBH (see above) MONUA Since July 1997 United Nations Observer Mission in Angola Strength: 2,795 Fatalities: 4 Budget appropriation July 1997–June 1998: $155 million (gross)	MIPONUH Since December 1997 United Nations Civilian Police Mission in Haiti Authorized strength: 300 Estimated cost for six months: about $14 million As of October 1997, revised budget estimates for the period July 1, 1997–June 30, 1998 amounted to $20.6 million (gross) and related to maintaining MIPONUH's predecessor missions, UNSMIH (ended June 1997) and UNTMIH (August–November 1997) and to the liquidation of UNTMIH

Source: United Nations Department of Public Information in consultation with the Department of Peacekeeping Operations and the Peacekeeping Financing Division (OPPBA), DPI/1634/Rev. 7, December 1997. Used by permission of the United Nations.

Note: UNTSO and UNMOGIP are funded from the UN's regular budget. Costs to the UN of the 13 other current operations are financed from their own separate accounts on the basis of legally binding assessments on all Member States. For these missions, the budget estimates and appropriations given above are for one year from July 1, 1997 to June 30, 1998, reflecting the financial periods established by the General Assembly in its resolution A/49/233A, unless otherwise indicated.

a. Strength figures include military and/or civilian police personnal on November 30, 1997.

b. Fatalities figures include military, civilian police, and civilian international and local personnel as of November 30, 1997.

    7. use force only in self-defense, and

    8. be supplied and administered under UN arrangements. (UN 1973)

Perhaps the most important conditions here are the nonuse of force except in self-defense and the consent of the parties to station a force. UNIFIL (Lebanon) and ONUC (Congo) have become classic examples of missions where several of the basic conditions were not met and where success was at best questionable.

It has been argued that it is not the use of force per se that is the problem, but the unsuccessful use of force where the intervention does not receive the support of the majority of the population. A guideline for success may be that force can only be effectively used to restrain a very small minority of the population (including a nonmajority-supported leadership) when that minority is violating agreed-upon norms, as is the case with the domestic use of police in participatory democratic societies.

The emphasis by some analysts on containing violence rather than resolving conflict has led others to conclude that either (1) peacekeeping is simply pacification, or (2) peacekeeping, by forestalling any "hurting stalemate," may remove any incentives for the parties to negotiate or resolve their conflict. Cyprus is often the classic case mentioned.

Probably the most important factor in success, then, is not the mission on the ground but whether it is accompanied by sufficient efforts to facilitate the resolution of the underlying conflict that led to the violence. While participants in the actual operation may do harm, if they are unskilled, partial, badly commanded, or without sufficient resources or information, they cannot resolve the underlying conflict on their own.

Reducing violence and helping resolve conflict are both important criteria for success, but the cost factor must also be considered. Conflict is costly in terms of time, money, and opportunities forgone; and violent conflict is even more costly, because it takes lives as well. Where conflicts remain unresolved, there is the constant potential for the resumption of violence. The cost of simply maintaining order without resolving underlying conflicts may not be worth it, either to the local or the international community. Perhaps a peacekeeping force that keeps order in a society for longer than a generation without resolving conflicts should be considered not a success but a failure. Perhaps a peacekeeping force, to provide real security, must always have both elements: the restraint of violence *and* assistance in the resolution of conflicts.

Other factors, such as leadership, organization, and communication, are also important. Adequate logistical support is critical. Secretary-General Boutros-Ghali (1995) makes particular note of this in *An Agenda for Peace*. Adequate and timely financing is also crucial to the success of peacekeeping or any organizational mission. Where peacekeeping forces are in considerable arrears, as many are, this constrains their effectiveness both logistically and politically. Perhaps financial pledges might be used as an additional gauge of the degree of consensus for an operation, with operations not being authorized unless there is clear financial support.

It is not certain that military personnel are the best peacekeepers. Michael Harbottle made the point that one of the lessons of UNEF, ONUC, and UNFICYP was that the professional soldier was no better as a peacekeeper than a volunteer. Volunteers are more likely to want to be there, and because they come from all walks of life, they may be more likely to have a common bond with those in the communities in which they are deployed (Harbottle 1971). Among the factors critical to the success of an operation are the attitudes of the members of the force, their ability to be sensitive to cultural differences, and their ability to solve problems and to facilitate the resolution of conflicts. Increasingly, the composition of peacekeeping missions is more and more diverse, including not only military personnel and civilian police, but administrators, observers, and other civilian personnel.

Civilians trained in the nonviolent resolution of conflict may have an increasingly important role to play in international peacekeeping, especially perhaps in the context of community-wide conflict. Nongovernmental nonviolent peacekeeping forces organized on the Gandhian *shanti sena* model have sent volunteers into situations of violence to monitor human rights or border crossings (see Desai 1972). Peace Brigades International is one of these, with an international board from all continents. Originally the World Peace Brigade, it was founded in Beirut, Lebanon, on January 1, 1962, and has sent transnational teams of observers to conflicts in Africa, Southeast Asia, and Central America. The conflict between the goals of reconciliation through mediation (which stresses impartiality) and confrontation in the name of justice (which tends to stress partisan support of one side of the struggle) has been a part of the strategy debate within Peace Brigades International.

A similar but later group, Witness for Peace, has operated both as a nonviolent witness and as a tripwire and communications device to keep U.S. citizens aware of what the U.S. and other governments are doing or planning in Central America. It has sent observers as support for local communities that are working for development and human rights in the face of repressive governments. As nongovernmental organizations become increasingly involved in day-to-day policymaking and administration in the United Nations, perhaps there is a role that trained, unarmed civilians skilled in nonviolent action and conflict resolution can play to improve the UN's peacemaking and peacekeeping capacity on the ground.

■ CONCLUSION

Perhaps the dilemmas the UN faces would not be so difficult if the international system were not so biased toward the utility of violent force. The efficacy of violence is a myth: at best, only some win—and at the great expense of others. Because violent force does not make peace or justice, its use by the UN would not seem to be the most cost-effective use of its limited resources.

It may well be that the situation the UN is in with respect to funding and general support may, in contrast to popular thinking, be due to the overuse and misuse of violent force rather than to its inability to bring about international (and subnational) peace and security. The Nobel Peace Prize was awarded after a period of many small, less violent peacekeeping missions. Criticism has come following the UN missions in Somalia and the former Yugoslavia. Perhaps it is time to have a more pointed debate on the merits of long- versus short-term, and more violent versus less violent, approaches to international peace, security, and justice.

If the international system is really to move in the direction of international peace, security, and justice, it must overcome the myth of the efficacy of violence and take on the more difficult task of creating security cooperatively. Less violent, longer-term means for providing international peace and security, such as peacemaking and peace-building, may hold much more promise than traditional enforcement—or even peacekeeping.

■ QUESTIONS

1. How have the concepts of peace and security evolved?

2. Do you think peacekeeping and peacemaking will be more attainable now that the Cold War is over?

3. Is it in the interest of the United States to cooperate with the United Nations in its peacekeeping and peacemaking efforts?

4. Is the world more secure now that the Cold War is over?

5. Should the United Nations have its own military force? How might this raise concerns over state sovereignty?

■ SUGGESTED READINGS

Boutros-Ghali, Boutros (1992) *An Agenda for Peace.* Second edition. New York: United Nations.
Harbottle, Michael (1971) *The Blue Berets.* London: Leo Cooper.
Higgins, Rosalyn (1996) *United Nations Peacekeeping.* 4 vols. London: Oxford University Press.
International Peace Academy (1984) *Peacekeeper's Handbook.* New York: Pergamon Press.
UN Department of Public Information (UNDPI) (1996) *The Blue Helmets: A Review of United Nations Peace-keeping.* Third edition. UN: UNDPI.
Weiss, Thomas G., ed. (1993) *Collective Security in a Changing World.* Boulder, CO: Lynne Rienner Publishers.
White, N. D. (1990) *The United Nations and the Maintenance of International Peace and Security.* New York: Manchester University Press.
Wiseman, Henry (1983) "United Nations Peacekeeping: An Historical Overview." In Henry Wiseman, ed. *Peacekeeping, Appraisals and Proposals.* New York: Pergamon Press.

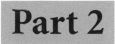

Part 2

The Global Economy

6

Controversies in International Trade

Bruce E. Moon

International trade is often treated purely as an economic matter that can and should be divorced from politics. That is a mistake, because trade not only shapes our economy but also determines the kind of world in which we live. The far-reaching consequences of trade pose fundamental choices for all of us. Certainly citizens must understand those consequences before judging the inherently controversial issues that arise over trade policy. More than that, we cannot even make sound consumer decisions without weighing carefully the consequences of our own behavior.

■ THE CASE FOR TRADE

The motives that generate international trade are familiar. Consumers who buy foreign products that are better or cheaper than domestic ones increase their material standard of living. Producers who sell their products for a higher price or in greater quantity abroad than at home increase their profit and wealth.

Most policymakers believe that governments should welcome trade because it provides benefits for the nation and the global economy as well as for the individual. Exports produce jobs for workers, profits for corporations, and revenues that can be used to purchase imports. Imports increase the welfare (well-being) of citizens because they can acquire more for their money as well as obtain products that are not available from domestic sources. The stronger economy that follows can fuel increasing power and prestige for the nation as a whole. Further, the resultant interdependence and shared prosperity among countries can strengthen global cooperation and maintain international peace.

Considerable evidence supports the view that trade improves productivity, consumption, and therefore material standard of living (Moon 1998). Trade successes have generated spurts of national growth, most notably in East Asia. The global economy has grown most rapidly during periods of trade expansion, especially after World War II, and has slowed when trade levels have fallen, especially during the Great Depression of the 1920s and 1930s. Periods of international peace have also coincided with trade-induced growth, while war has followed declines in trade and prosperity.

The private benefits of trade have led individual consumers and producers to embrace it with zeal for the last half-century. As a result, trade has assumed a much greater role in almost all nations, with exports now constituting about a quarter of the economy in most countries and well over half in many. Even in the United States, which is less reliant on trade than virtually any other country in the world, the export sector is now about 10 percent of GNP (gross national product, defined as the total of goods and services produced by a country's citizens in a given year). Smaller nations, especially those of the third world, must engage in more trade because they can neither supply all their own needs nor provide a market sizable enough for their own industries.

Governments have not prevented this growth in trade levels; in fact, at least since World War II, most have encouraged and promoted it. All but a handful of nations now rely so heavily on jobs in the export sector and on foreign products to meet domestic needs that discontinuing trade is no longer an option. To attempt it would require a vast restructuring that would entail huge economic losses and massive social change. Furthermore, according to the "liberal" trade theory accepted by most economists, governments have no compelling reason to interfere with the private markets that achieve such benefits. Liberal economic theory is not to be confused with the ambiguous way that the term *liberal* is applied in U.S. politics. Liberalism, as used throughout this chapter, opposes government interference with the market.

From its roots in the work of Scottish political economist Adam Smith (1723–1790) and English economist David Ricardo (1772–1823), this liberal perspective has emphasized that international trade can benefit all nations simultaneously, without requiring governmental involvement (Smith 1910). According to Ricardo's theory of comparative advantage (Ricardo 1981), no nation need lose in order for another to win, because trade allows total global production to rise. The key to creating these gains from trade is the efficient allocation of resources whereby each nation specializes by producing goods in which it has a comparative advantage. For example, a nation with especially fertile farmland and a favorable climate could produce food much more cheaply than a country that lacked this comparative advantage. Similarly, a nation with efficient manufacturing facilities could produce clothing more cheaply than one better suited to raising crops. If countries each specialized in the goods they produced efficiently, they could trade their excess production to one another and both would be better off, because trade allows each nation to apply its resources to their most efficient use. No action by governments is required to bring about this trade, however, since profit-motivated investors will see to it that producers specialize in the goods in which they have a comparative advantage, and consumers will naturally purchase the best or cheapest products. Thus, liberal theory concludes that international trade conducted by private actors free of government control will maximize global welfare.

Though trade levels have grown massively in the two centuries since Adam Smith, government regulation of trade has varied from tightly constraining to relatively permissive. However, no government has followed the advice of economic theorists to refrain from controlling trade altogether. This curious disparity between accepted economic theory and established political practice results in part from concerns about whether trade is always beneficial and in part from doubts about whether government action is required to bring about the benefits of trade. The most notable focal point of government anxiety surrounds the disruptive impact of the trade deficits that often arise in trading nations. A trade deficit exists when a nation's imports are greater than its exports, meaning that its residents buy more from other nations than it sells to them.

■ TRADE DEFICITS

Most states try to control trade in order to minimize its potentially negative consequences; at the same time, they want to capture the benefits claimed for it by liberal theory. In walking this tightrope, governments have sought trade expansion just as economic liberalism would recommend, but they also have been heavily influenced by a dissenting body of thought known as *mercantilism*. Mercantilism originated with the "mercantile" policy (i.e., commercial or trade policy) of European nations, especially England, from the sixteenth century to the middle of the nineteenth. While mercantilism does not oppose trade, it holds that governments must regulate it in order for trade to advance various aspects of the national interest. The long-term aspirations of mercantilists go beyond immediate consumption gains to include national self-sufficiency, the vitality of key industries, and a powerful state in foreign policy. The immediate goal—and often the focal point of trade policy in mercantilist states—is to avoid incurring a deficit in the balance of trade.

Mercantilists observe that the rosy evaluation of trade advanced by Smith and Ricardo was predicated on their expectation that any given nation's imports would more or less balance its exports. However, despite the efforts of nations to avoid potentially damaging imbalances, trade deficits are now considered commonplace. For example, the United States has run a substantial trade deficit for more than two decades, with imports surpassing exports by over $100 billion annually in recent years. The consequences of these deficits are complex and controversial.

If a nation's imports are greater than its exports, it follows that national consumption must exceed its production. One might ask how anyone could object to an arrangement that allows a nation to consume more than it produces: after all, a trade deficit amounting to more than $500 per person annually allows U.S. citizens to enjoy a higher standard of living than would otherwise be possible. The answer lies in recognizing that such a situation must have adverse repercussions (especially in the future), just as individuals cannot continue to spend more than they earn without eventually suffering detrimental consequences.

Mercantilists fear that excessive imports permit foreigners to enjoy employment and profits from production that might otherwise benefit citizens of the home country. On the face of it, a trade deficit of over $100 billion per year would appear to transfer millions of jobs out of the U.S. economy. For example, high levels of unemployment in Detroit and low levels of unemployment in Tokyo have been ascribed partially to the massive sales of Japanese cars in the United States.

Liberals correctly note, however, that trade imbalances cannot be evaluated so easily, in part because they trigger complex flows of money that are also unbalanced. For example, the U.S. trade deficit means that more money flows *out* of the United States economy in the form of dollars to pay for imports than flows back *into* the economy through payments for U.S. goods purchased by foreigners. The consequences of the trade deficit depend in large part on what happens to the excess dollars that would appear to be piling up abroad.

In practice, most of those dollars have found their way back into the U.S. economy, because foreigners have used them to purchase U.S. Treasury bonds and real estate and to finance the takeover of U.S. businesses as well as make new investments in the United States. Such capital flows can offset a trade deficit and render it harmless in the short run, but they only postpone the inevitable need to balance production and consumption. For example, foreigners now hold over $1,000 billion in U.S. Treasury debt, about 30 percent of all treasuries outstanding. Someday that debt must be repaid. In the meantime, these investments produce income for foreigners. For example, foreigners now receive about $70 billion in interest payments annually just from the U.S. federal government, an amount that constitutes about half the current federal budget deficit and is growing rapidly (Forsyth 1996).

Economists disagree about whether these developments ought to raise alarm. After all, the willingness of foreigners to invest in the United States surely is an indication of confidence in the strength of the U.S. economy. More generally, as Chapter 7 shows, capital flows can be beneficial to the economy and its future. Indeed, foreign capital is an essential ingredient to development in many third world countries. Whether capital inflows produce effects that are, on balance, positive or negative depends heavily on the source of the capital, the terms on which it is acquired, and especially on the uses to which it is put.

International currency markets have sent a cautionary signal that the persistent U.S. trade deficit is eroding the confidence of foreign investors. Even with massive inflows of investment from abroad, the demand for dollars by foreigners—to purchase products or investments from the United States—has been smaller than the supply of dollars created by Americans purchasing foreign products and investments. As a result, the value of the dollar, once equivalent to 360 Japanese yen (¥360), declined to under ¥80 in early 1995 (before recovering to ¥125 by early 1997). These currency fluctuations affect the purchasing power of the dollar. For example, the ¥1,000,000 cost of a Japanese automobile would translate into a dollar price of about U.S.$2,800 at the exchange rate of 360 yen per dollar, but would require more than U.S.$10,000 when the dollar fell below ¥100.

Balance of trade deficits tend to lead to such currency declines and to future price increases and snowballing debt. Thus, a trade deficit provides immediate benefits, but also implies that future consumption will be reduced and the standard of living for future generations will fall. Whether this is viewed as a good thing or not is a subjective matter. Thus, nations vary greatly in their tolerance for trade deficits, but almost all try to minimize or avoid them altogether. Indeed, many nations even seek to accumulate a trade surplus, in which exports exceed imports.

■ OPTIONS IN TRADE POLICY

To achieve their desired trade balance, nations often combine two mercantilist approaches. They may emphasize the expansion of exports through a strategy known as industrial policy. More commonly, they emphasize minimizing imports, a stance known generally as *protectionism* (Fallows 1993).

Protectionist policies include many forms of import restrictions designed to limit the purchase of goods from abroad so that the domestic import-competing industries can capture a larger share of the market. The simplest of these are quotas, government restrictions that place a fixed limit on the quantity or value of goods that can be imported. This is usually accomplished by requiring that importers obtain import licenses, which are strictly rationed by governments. The usual effect of a quota is to raise the domestic price of the commodity by limiting the number of lower-priced products that can be imported. This allows domestic producers to gain a larger market share but requires that consumers pay a higher price. Both domestic and foreign producers are able to maintain higher prices because of this artificial restriction of supply.

The most traditional import barriers are tariffs (or import duties), which come in two forms. Most are ad valorem, calculated as a percentage of the value of the good imported. U.S. ad valorem tariffs average about 4 percent today, about the same as in most developed nations. Specific tariffs are applied to a particular quantity of an imported good; for example, the United States imposes a fixed tariff on every barrel of imported oil.

Many forms of protectionism have increased in the past decade as governments have responded to the pleas of industries threatened by foreign competition. But tariffs are no longer the main form of protectionism. In fact, declining from their peak in the 1930s, tariff levels throughout the world are generally very low. In the United States, the average tariff rate reached a modern high of 59 percent in 1932 under what has been called "a remarkably irresponsible tariff law," the Smoot-Hawley Act, which has been widely credited with triggering a spiral of restrictions by other nations that helped plunge the global economy into the Great Depression of the 1930s. It was reduced to 25 percent after World War II and declined to about 5 percent after the Tokyo Round of trade negotiations concluded in 1979.

In place of tariffs, a variety of nontariff barriers (NTBs) have arisen, especially voluntary export restraints (VERs). In the most famous case of VERs, Japanese automakers "voluntarily" agreed to limit exports to the United States in 1981. (Had Japan refused, a quota that would have been more damaging to Japanese automakers would have been imposed.) The Federal Trade Commission (FTC) has estimated that the higher prices for autos that resulted cost U.S. consumers about $1 billion per year. Not only does the restricted supply of Japanese autos cause their prices to rise because of the artificial shortage, but it also enables U.S. manufacturers to maintain higher prices in the absence of this competition.

A favorable trade balance—or the elimination of an unfavorable one—also can be sought through an industrial policy that promotes exports. The simplest technique is a *direct export subsidy,* in which the government pays a domestic firm for each good exported, so that it can compete with foreign firms that otherwise would have a cost advantage. Such a policy has at least three motivations. First, by increasing production in the chosen industry, it reduces the unemployment rate. Second, by enabling firms to gain a greater share of foreign markets, it gives them greater leverage to increase prices (and profits) in the future. Third, increasing exports will improve the balance of trade and avoid the problems of trade deficits.

Liberals are by no means indifferent to the dangers of trade deficits, but they argue that most mercantilist cures are worse than the disease. When mercantilist policies affect prices, they automatically create winners and losers and in the process engender political controversies. For example, because the revenue to pay for that subsidy must be raised through taxes, the domestic consumer has to pay higher taxes. As noted above, protectionism also harms the consumer by raising prices even while it benefits domestic firms that compete against imports.

If mercantilist policies are controversial in the nations that enact them, they are met with even greater hostility by the nations with which they trade. For example, Japanese protectionism prevents U.S. firms from competing for the lucrative Japanese market, while Japanese export promotion policies place U.S. firms at a disadvantage even in the U.S. market. The U.S. steel industry has been particularly outspoken in its denunciation of steel imported from foreign firms that are heavily subsidized by their governments. They contend that U.S. jobs and U.S. profits are being undercut by this unfair competition.

The United States has generally preferred to maintain a desirable volume and balance of trade by inducing other nations to lower their trade barriers rather than by erecting its own. The United States has undertaken direct bilateral negotiations to change the policies of other nations, especially Japan, and has spearheaded efforts to create and maintain global institutions that facilitate trade.

The Bretton Woods trade and monetary regime, created under the leadership of the United States at the end of World War II, is the outstanding example of such institutions at the global level. The Bretton Woods regime, centered on the institutions of the General Agreement on Tariffs and Trade (GATT) and the International Monetary Fund (IMF), has governed international trade and finance for nearly fifty years, though it has evolved and changed markedly in that time. The IMF has sought to expand trade by guaranteeing stability in monetary affairs and by providing mechanisms to finance imports and adjust trade imbalances. Since 1946, the GATT has convened eight major negotiating sessions in which nations exchange reductions in tariffs and other trade barriers, with the result that global trade has increased dramatically. The most recent session, the Uruguay Round completed in 1994, also created the World Trade Organization (WTO) to provide a setting for the GATT's 123-nation membership to resolve trade disputes peacefully. The European Union (EU), which represents the most extensive and most successful effort to achieve free trade at the regional level, also inspired the North American Free Trade Agreement (NAFTA), designed to expand trade among the United States, Canada, and Mexico.

■ THE MULTIPLE CONSEQUENCES OF TRADE

As nations choose among policy options, they must acknowledge liberal theory's contention that free trade allows the market to efficiently allocate resources and thus to maximize global and national consumption. Neither can the desire of individual consumers and producers to participate in trade be ignored. As our brief survey of the consequences of a trade deficit illustrates, however, the simplicity of individual motives conceals the complexity of the effects that trade has on others. As we are about to see, the dangers of trade deficits are only a small part of why governments almost universally restrict trade.

In fact, governments seek many outcomes from trade—full employment, long-term growth, economic stability, social harmony, power, security, and friendly foreign relations—yet discover that these desirable outcomes are frequently incompatible with one another. Because free trade may achieve some goals but undermine others, governments that fail to heed the advice of economic theory need not be judged ignorant or corrupt. (Instead, they recognize a governmental responsibility to cope with all of trade's consequences, not only those addressed by liberal trade theory.) For example, while trade affects the prices of individual products, global markets also influence which individuals and nations accumulate wealth and political power. Trade determines who will be employed and at what wage. It determines what natural resources will be used and at what environmental cost. It shapes opportunities and constraints in foreign policy.

Because trade affects such a broad range of social outcomes, conflict among alternative goals and values is inevitable. As a result, both individuals and governments must face dilemmas that involve the multiple consequences of trade, the multiple goals of national policy, and the multiple values that compete for dominance in shaping our behavior (Moon 1996).

■ THE DISTRIBUTIONAL EFFECTS OF TRADE

Many of these dilemmas stem from the sizable effect that international trade has on the distribution of income and wealth among individuals, groups, and nations. Simply put, some gain material benefits from trade while others lose. Thus, to choose one trade policy and reject others is simultaneously a choice of one income distribution over another. As a result, trade is inevitably politicized: each group pressures the government to adopt a trade policy from which they expect to benefit.

The most visible distributional effects occur because trade policy often protects or promotes one industry or sector of the economy at the expense of others. For example, tariffs on imported steel protect the domestic steel industry by making foreign-produced steel more expensive, but they also harm domestic automakers who must pay higher prices for the steel they use. As in this case—where car buyers face higher prices—most barriers to trade benefit some sector of the economy at the expense of consumers, a point always emphasized by proponents of free trade.

Trade policy also benefits some classes and regions at the expense of others, a point more often emphasized by those who favor greater governmental control. For example, the elimination of trade barriers between the United States and Mexico under NAFTA terms forces some U.S. manufacturing workers into direct competition with Mexican workers, who earn a markedly lower wage. Unless U.S. wage rates decline, production may shift to Mexico and U.S. jobs will be lost. However, if that labor competition drives down wage rates in the United States, the profits earned by the U.S. business owners might be maintained at the expense of the standard of living of workers in those industries. The losses from such wage competition will be greatest for workers in high-wage countries employed in industries that can move either their products or their production facilities most easily across national boundaries. Others, particularly more affluent professionals who face less direct competition from abroad (such as doctors, lawyers, and university professors), stand to gain from trade because it lowers prices on the goods they consume.

Proponents of free trade tend to deemphasize distributional effects and instead focus on the impact of trade on the economy as a whole. That is partly because liberal theory contends that free trade does not decrease employment but only shifts it from an inefficient sector to one in which a nation has a comparative advantage. For example, U.S. workers losing their jobs to Mexican imports should eventually find employment in industries that export to Mexico. Proponents of free trade insist that it is far better to tolerate these "transition costs"—the short-term dislocations and distributional effects—than to protect an inefficient industry.

Because these distributional consequences have such obvious political implications, however, the state is much more attentive to them than economic theorists are. That is one reason all governments control trade to one degree or another. Of course, that does not mean that they do so wisely or fairly, in part because their decisions are shaped by patterns of representation among the constituencies whose material interests are affected by trade policy. In general, workers tend to be underrepresented, which is why trade policies so often encourage trade built on wage rates that enrich business owners but constrain the opportunities for workers. Similarly, the economic structures created by trade patterns can produce just as great a distributional inequity between genders as between classes, sectors, or regions. As Chapter 10 describes, political representation of women at all levels of decisionmaking is very poor. Finally, the economic activities shaped by trade policies tend to affect current generations very differently from future ones—and the latter are seldom represented at all.

■ THE VALUES DILEMMA

These distributional effects pose challenging trade-offs among competing values. For example, the effects of NAFTA were predicted to include somewhat lower prices for U.S. consumers, but also job loss or wage reduction for some unskilled U.S. workers. The positions taken on this issue by most individuals, however, did not hinge on their own material interests; few could confidently foresee any personal impact of NAFTA since the gains were estimated at well under 1 percent of GNP, and job losses were not expected to exceed a few hundred thousand in a labor force of more than 100 million. However, the choice among competing values was plain: NAFTA meant gains in wealth but also greater inequality and insecurity for workers. Some citizens acceded to the judgment of liberal theory that the country as a whole would be better off with freer trade, while others identified with the plight of workers, who were more skeptical of liberal theory simply because for them the stakes were so much higher. After all, it is far easier for a theorist to move a column of figures than it is for a worker who has devoted his life to farming to pack up and move to a strange town in the hopes that he might find a job in an unfamiliar industry that requires skills he does not possess. In the final analysis, NAFTA became a referendum on what kind of society people wished to live in. The decision was quintessentially American: one of greater wealth but also greater inequality and insecurity.

Of course, other distributional effects gave rise to other value choices as well. Since the gains from NAFTA were expected to be greater for Mexico than for the United States, the conscientious citizen would also weigh whether it is better to help Mexican workers because they are poorer or to protect U.S. workers because they are U.S. citizens. As Chapter 8 implies, such issues of inequality in poor societies can translate directly into questions of life or death. As a result, the importance of trade policy, which has such a powerful impact on the distribution of gains and losses, is heightened in poor, dependent nations (especially in an economy with half of GNP related to trade).

Perhaps the most challenging value trade-offs concern the trade policies that shift gains and losses from one time period to another. Such "intergenerational" effects arise from a variety of trade issues. For example, we have seen that the U.S. trade deficit, like any form of debt, represents an immediate increase in consumption but a postponement of its costs. Interestingly, the Japanese industrial policy of export promotion fosters a trade surplus that reinforces the U.S. preference for immediate gratification, while it produces the opposite effect in Japan. The subsidies the Japanese government pays to Japanese exporters make their products cheaper abroad, so consumers in countries like the United States benefit. At the same time, those subsidies require Japanese citizens to pay both higher prices and higher taxes. However, the sacrifices of Japan's current generation may benefit future ones if this subsidy eventually transforms an "infant industry" into a powerful enterprise that can repay the subsidies through cheaper prices or greater employment. Meanwhile, if the subsidies drive U.S. firms out of business, future generations of U.S. citizens may suffer losses of employment opportunities and higher prices. It is interesting to speculate as to why U.S. policies so frequently differ from Japanese ones when distributional effects pose the values dilemma of whether it is better to sacrifice now for the future or to leave future generations to solve their own problems.

The values dilemma encompasses much more than just an alternative angle on distributional effects, however (Polanyi 1944). The debate over "competitiveness," which began with the efforts by U.S. businesses to lower their production costs in order to compete with foreign firms, illustrates how trade may imply a compromise of other societal values. Companies often find that government policies make it difficult for them to lower their labor costs. Lower wages could be paid if the minimum wage were eliminated and collective bargaining and labor unions were outlawed. The abolition of seniority systems and age discrimination laws would enable companies to terminate workers when their efficiency declined (or at the whim of a boss). Eliminating pensions, health care, sick leave, workman's compensation for accidents, workplace safety regulations, and paid vacations and holidays would also lower company labor costs. But such actions entail a compromise with very fundamental values, because most U.S. citizens believe that national policy should seek to make corporations more competitive in order to improve the lives of its citizens, not compromise citizen welfare in order to improve competitiveness. The government regulations that handicap U.S. business were designed to meet other legitimate national goals. Environmental regulations, for example, may add to production costs, but surely the pressures of international trade need not require that we abandon all other values. With respect to natural resources, trade has substantial impact on land use patterns and other aspects of the macroenvironment. Choosing between such alternative values is always difficult for a society, because reasonable people can differ in the priority they ascribe to such values. Still, agreements on such matters can usually be forged within societies, in part because values tend to be broadly, if not universally, shared.

Unfortunately, free trade forces U.S. firms burdened by these value choices to compete with firms operating in countries that may not share them. This situation creates a dilemma for U. S. consumers as well as for the U.S. government, because both have the capacity to ease the competitive pressures on U.S. firms if they choose to do so. But that choice involves a difficult balance between economic interests and other values. For example, continuing to trade with nations that permit shabby treatment of workers—or even outright human rights abuses—poses a difficult moral dilemma. As Chapter 4 documents, foreign governments have often declared their opposition to human rights abuses but have seldom supported their rhetoric with actions that effectively curtailed the practice. In fact, the implementation of the policy to pressure foreign governments on behalf of a normative stance has been left to consumers, who have unwittingly answered key questions daily: Should we purchase cheap foreign goods like clothing and textiles even though they may have been made with child labor—or even slave labor? If we do, can we really blame U.S. companies for moving their production facilities abroad?

Where values are concerned, of course, we cannot expect everyone to agree with the choices we ourselves might make. As Chapter 11 describes, child labor remains a key source of comparative advantage for many countries in several industries prominent in international trade. We cannot expect them to give up easily a practice that is a major component of their domestic economy and that is more offensive to us than to them. If trade competitors do not share our values, it may prove difficult to maintain them ourselves—unless we restrict trade, accept deficits, or design state policies to alleviate the most dire consequences. After all, it is hard to see how U.S. textile producers can compete with the sweatshops of Asia without creating sweatshops in New York, a dilemma in which even media figures such as Kathie Lee Gifford have recently become involved.

Further, if we restrict trade because we oppose child labor or rain forest destruction, how can we object when other countries ban the sale of U.S. products because they violate *their* values—a criticism frequently made of U.S. rock music, Hollywood films, McDonald's hamburgers, and other symbols of U.S. cultural imperialism? At issue is the tension between maintaining fair competition among firms in different countries—which is essential to sustaining the international trading system—and maintaining the cultural and political differences among nations—which is central to the national sovereignty and autonomy of the modern state system.

■ FAIR COMPETITION AND NATIONAL AUTONOMY

Fair competition in trade requires at least implicit cooperation between governments, because no nation can export unless some other nation imports. However, while nations usually encourage the exports on which they rely for jobs, for profits, and for the limitation of balance of trade deficits, they are usually less enthusiastic about welcoming imports. Fortunately, almost all nations acknowledge, in principle, the obligation to permit the sale of foreign products within their borders, if only because they fear that excessive protectionism of their own market will encourage other nations to protect theirs.

Still, disputes over trade barriers are common, because—in practice—governments have many compelling motives for enacting policies that affect trade. Often one nation defends its policy as a rightful exercise of national sovereignty, while another challenges it as an unfair barrier to trade. Ideally, such disagreements have been settled by appeal to the GATT or, more recently, to the WTO, whose new dispute resolution panel hears trade disputes and determines whether national behavior is consistent with international rules. While the U.S. administration strongly supported the creation of the WTO to prevent trade violations by other nations, a surprising variety of U.S. groups opposed its ratification because it might encroach on national sovereignty. Environmental groups such as Friends of the Earth, Greenpeace, and the Sierra Club were joined not only by consumer advocates like Ralph Nader, but also by conservatives such as Ross Perot, Pat Buchanan, and Jesse Helms, who feared that a WTO panel could rule that various U.S. government policies constituted unfair trade practices, even though they were designed to pursue values utterly unrelated to trade. For example, EU automakers have challenged the U.S. law that establishes standards for auto emissions and fuel economy. Buchanan said, "WTO means putting America's trade under foreign bureaucrats who will meet in secret to demand changes in United States laws. . . . WTO tramples all over American sovereignty and states' rights" (Dodge 1994). Because the WTO could not force a change in U.S. law, GATT director-general Peter Sutherland called this position "errant nonsense" (Tumulty 1994), but the WTO could impose sanctions or authorize an offended nation to withdraw trade concessions as compensation for the injury.

The most dramatic example occurred in the 1994 case known as "GATTzilla versus Flipper," in which a GATT tribunal ruled in favor of a complaint brought by the EU on behalf of European tuna processors who buy tuna from Mexico and other countries that use purse seine nets. The United States boycotts tuna caught in that way because the procedure also kills large numbers of dolphins; but this value is not universally shared by other nations. In fact, the GATT ruled that the U.S. law was an illegal barrier to trade because it discriminates against the fishing fleets of nations that use this technique.

Regional agreements cannot avoid this clash between fair competition in trade and national autonomy. The first trade dispute under NAFTA involved a challenge by the United States to regulations under Canada's

Fisheries Act established to promote conservation of herring and salmon stocks in Canada's Pacific Coast waters. Soon thereafter the Canadian government challenged U.S. Environmental Protection Agency (EPA) regulations that require the phasing out of asbestos, a carcinogen no longer permitted as a building material in the United States (Cavanaugh et al. 1992).

Similarly, critics of the EU worry that its "leveling" of the playing field for trade competition also threatens to level cultural and political differences among nations. Denmark, for example, found that free trade made it impossible to maintain a sales tax rate higher than neighboring Germany's, because Danish citizens could simply evade the tax by purchasing goods in Germany and bringing them across the border duty free. Competitiveness pressures also make it difficult for a nation to adopt policies that impose costs on business when low trade barriers force firms to compete with those in other countries that do not bear such burdens. For example, French firms demand a "level playing field" in competing with Spanish firms whenever the French government mandates employee benefits, health and safety rules, or environmental regulations more costly than those in Spain. In fact, free trade tends to harmonize many national policies.

Thus, some trade barriers are designed to protect unique aspects of the economic, social, and political life of nations, especially when trade affects cultural matters of symbolic importance. For example, France imposes limits on the percentage of television programming that can originate abroad, allegedly in defense of French language and custom. The obvious target of these restrictions, U.S. producers of movies and youth-oriented music, contend that the French are simply protecting their own inefficient entertainment industry. Indeed, Hollywood sees "gangsta" rap and *Baywatch* as valuable export commodities that deserve the same legal protection abroad that the foreign television sets and CD players that display these images receive in the United States. Others see them as an example of cultural imperialism, a threat to moral values, or even an outright assault on national security. Can it be long before Colombia challenges U.S. drug laws as discriminating against marijuana while favoring Canadian whiskey?

■ FOREIGN POLICY CONSIDERATIONS IN INTERNATIONAL TRADE

Some of the most challenging value choices concern the effect of trade on the foreign policy goals pursued by states, such as power and peace as well as national autonomy. For example, policymakers have long been aware that trade has two deep, if contradictory, effects on national security. On the one hand, trade contributes to national prosperity, which increases national power and enhances security. On the other hand, it has the same effect on a nation's trade partner, which could become a political or even military rival. The resulting ambivalent attitude is torn between the vision of states cooperating for economic gain and the recognition that they also use trade to compete for political power.

While a market perspective sees neighboring nations as potential customers, the state must also see them as potential enemies. As a result, the state must not only consider the absolute gains it receives from trade, but also must weigh those gains in relative terms, perhaps even avoiding trade that would more be advantageous to its potential enemies than to itself. For this reason, states have always been attentive to the distribution of the gains from trade and selective about their trade partners, frequently encouraging trade with some nations and discouraging or even banning it with others.

While understandable, such policies create competitive struggles for markets, raw materials, and investment outlets, which sometimes can lead to open conflict. In fact, U.S. president Franklin Roosevelt's secretary of state Cordell Hull even went so far as to contend that "bitter trade rivalries" were the chief cause of World War I and a substantial contributor to the outbreak of World War II. Both were precipitated by discriminatory trade policies in which different quotas or duties were imposed on the products of different nations. Hull, who believed that free multilateral trade would build bridges rather than create chasms between peoples and nations, thus championed the nondiscrimination principle, which became embodied in Article I's most-favored-nation (MFN) clause of GATT. As we saw above, however, it also became the grounds for challenging as discriminatory some of the laws that manifest national autonomy.

A similar belief in the efficacy of free trade as a guarantor of peace was an important motivation for the initiation of the regional integration in Europe that eventually created the European Union. This process was launched in 1951 with the founding of the European Coal and Steel Community (ECSC), which internationalized an industry that was key not only for the economies of the six nations involved, but also for their war-making potential. With production facilities scattered among different countries, each became dependent on the others to provide both demand for the final product and part of the supply capacity. This arrangement fulfilled the liberal dream of an interdependence that would prevent war by making it suicidal. In fact, the ECSC was an innovative form of peace treaty, designed, in the words of Robert Schuman, to "make it plain that any war between France and Germany becomes, not merely unthinkable, but materially impossible" (Pomfret 1988: 75).

Nations differ in their response to this interdependence, however. Canadian fear of the economic dominance of the United States has long colored relations between the two, and Britain remains ambivalent about closer ties with Europe because of fear that national autonomy must be sacrificed to achieve them. In the cases of the EU and Bretton Woods, however, policymakers saw several ways that an institutionalized liberal trading system could promote peace among nations. The growth of global institutions could weaken the hold of nationalism and mediate conflict between nations. Trade-induced contact could break down nationalistic hostility among societies. Multilateralism (nondiscrimination) would tend to prevent grievances from developing among states. Interdependence could weaken nationalism, constrain armed conflict, and foster stability. The economic growth generated by trade could remove the desperation that leads nations to aggression.

Such institutions are helpful in reducing the conflict potential in trade relations, but they require leadership to overcome the inclination of most nations to retain their own trade barriers while inducing other countries to lower theirs. One dominant nation will usually have to subsidize the organizational costs and frequently offer side benefits in exchange for cooperation, such as the massive infusion of foreign aid provided to Europe by the United States under the Marshall Plan in the late 1940s.

Maintaining the capability to handle these leadership requirements has substantial costs for the United States. For example, U.S. expenditures for defense, which have been many times higher than those of nations with whom it competes since World War II, erode the competitiveness of U.S. business by requiring higher tax levels; they constrain the funds available to spend on other items that could enhance competitiveness; and they divert a substantial share of U.S. scientific and technological expertise into military innovation and away from commercial areas. (Ironically, much of that money has been spent directly on protecting the very nations against whom U.S. competitiveness has slipped, especially Germany, Japan, and Korea.) The trade-off between competitiveness and defense may be judged differently by different individuals, but it can be ignored by none. To give up global leadership or national security may be a wise choice, but it is not without costs of its own.

■ CONCLUSION:
CHOICES FOR NATIONS AND INDIVIDUALS

Few would deny the contention of liberal theory that trade permits a higher level of aggregate consumption than would be possible if consumers were prevented from purchasing foreign products. It is hard to imagine modern life without the benefits of trade. However, this aggregate economic effect tells only part of the story, because trade also carries with it important social and political implications. Trade shapes the distribution of income and wealth among individuals, affects the power of states and the relations among them, and constrains or enhances the ability of both individuals and nations to achieve goals built on other values. Thus, trade presents a dilemma for nations: no policy can avoid some of trade's negative consequences without also sacrificing its aggregate economic benefits. That is why most governments have sought to encompass elements of both liberalism and mercantilism in fashioning their trade policies. The same is true for individuals, because every day each individual must—explicitly or implicitly—assume a stance on the dilemmas we have identified. In turn, it forces individuals to consider some of the following questions—questions that require normative judgments as well as a keen understanding of the empirical consequences of trade. We must always remember to ask not only what trade policy will best achieve our goals, but also what should our goals be.

■ QUESTIONS

1. Are your views closer to those of a liberal or a mercantilist?

2. Is it patriotic to purchase domestic products? Why or why not?

3. Do we owe greater obligations to domestic workers and corporations than to foreign ones?

4. Should one purchase a product that is cheap even though it was made with slave labor or by workers deprived of human rights?

5. Should one choose transportation that requires the importation of foreign oil, knowing it encourages a costly U.S. military presence in the Middle East?

6. Should one lobby the government to restrict the sales of U.S. forestry products abroad because they compromise environmental concerns?

■ SUGGESTED READINGS

Fallows, James (1993) "How the World Works," *Atlantic Monthly*, December.
Moon, Bruce E. (1996) *Dilemmas of International Trade*. Boulder, CO: Westview Press.
———— (1998) "Exports, Outward-Oriented Development, and Economic Growth," *Political Research Quarterly* (March).
Polanyi, Karl (1944) *The Great Transformation*. New York: Farrar & Reinhart.
Ricardo, David (1981) *Works and Correspondence of David Ricardo: Principles of Political Economy and Taxation*. London: Cambridge University Press.
Smith, Adam (1910) *An Inquiry into the Nature and Causes of the Wealth of Nations*. London: J.M. Dutton.

International Capital Flows

Gerald W. Sazama

As humans have developed technology, our race has simultaneously developed economic systems. Primitive economic exchange occurred through the barter of handmade goods within a clan or between neighboring tribes. Modern economic exchange involves the transfer of simple products made with complex machinery, such as wheat, or of the complex machinery itself. These modern exchanges are financed by money and other sophisticated financial instruments. In finance our race has moved from simple forms of money like wampum beads and gold coins to paper money and checks, and then to international markets for investment funds and national currencies.

Like the growth in humans' ability to develop technology, the increasing sophistication of economic systems and financial instruments is wonderful. This sophistication makes possible the complex movement of products and resources first within single countries and now around the globe. Like the development of technology, however, the growth of international capital flows can also be used in horrible ways: for example, for selfish accumulation or for capital flight that dries up a nation's savings base, pushing it into a depression.

Topics discussed in this chapter are (1) international capital flows—what they are and how they work; (2) capital flows between the industrialized or more developed countries (MDCs), such as the Japanese building a Honda plant in Ohio, and the Americans having a Chase Manhattan

I am indebted to John Powelson of the University of Colorado for his assistance in gathering material and in writing some portions of this chapter.

branch bank in London; and (3) capital flows between the MDCs and the less developed countries (LDCs). These capital flows occur as foreign investment, debt, or foreign aid.

■ BASICS CONCEPTS OF INTERNATIONAL FINANCE

Three concepts central to international finance are capital theory, monetary theory, and exchange rates.

▨ Capital Theory

Economists divide capital into three types: physical, human, and financial. *Physical capital* is all the tools, machinery, and buildings we use to make products and services. At its core, this capital is saved labor, part of output that is not consumed but saved for future use. On its basic level, physical capital is the twelfth-century Native American not consuming all her corn crop so she has seeds to plant next spring. Even a computer is at its essence saved human labor, postponed consumption, embodied in a specific machine. *Human capital* is investing in ourselves. With training, creativity, discipline, and study, we gain the capacity to produce goods and services more efficiently. We then develop and combine these human abilities through institutions such as universities or private manufacturing corporations. *Financial capital* is the paper instruments through which society signals ownership rights over physical and organized human capital. Thus, a share of stock in General Motors gives its owner a claim on part of the physical capital and human capital organized within General Motors, and on the profits generated by this capital.

As these forms of capital grow more complex, it is helpful to keep in mind the basics of capital theory. True capital, both physical and human, is nothing more than saved labor. To invest we have to save and thus not consume part of our output; rather we must hold it to increase our future output. Paper, or financial capital, has no value in itself. It is useful because it facilitates exchange and movement of the physical and human capital. Paper, or financial capital, has little or no value if it represents ownership of a broken machine or building or a dysfunctional corporate organization.

▨ Monetary Theory

Money and banking. Originally trade occurred by bartering one good for another. If the candlestick maker, however, did not want the butcher's meat because he was a vegetarian, it would be difficult for the village butcher to get candles. Therefore, our invention of money was very useful to the growth of exchange. An early form of money was precious metals. But as trade grew among regions, it became inconvenient and dangerous to carry around these precious metals. So the gold was kept at the goldsmith's. The goldsmith then gave you a slip of paper (a note) that said you had gold on deposit with him—thus the birth of paper currency. Later, some wise person named Mr. Green wrote Mr. Black a slip of paper so that Mr. Black could go to Mr. Green's goldsmith and claim some of his precious metals—thus the birth of checks.

Next, the goldsmith realized that all his customers were not going to claim all their gold at the same time, so he could lend it out to a guy named Marco Polo, earning interest in the process. Mr. Polo would return to Italy with spaghetti and gun powder, repay the loan, and earn a profit in the process—thus banks create money "out of nothing"; nothing, that is, but trust that Marco Polo will repay his loan.

Banks, however, can create too much money—more than there are real goods and services to purchase—causing inflation. Inflation is bad for banks because their loans are repaid in money that is worth less than it was when they lent it out. Some banks could also put out too many loans and not be careful to whom they lent the money. When customers came back for their money at such banks there would be no assets left. These problems of inflation and imprudent banking practices led to the nineteenth-century financial panics and were instrumental in the U.S. savings and loan crisis in the 1980s. As a result of these financial panics, humans invented central banks to regulate commercial banks.

Flexible exchange rates and arbitrage. Basic to the growth of imports and exports is the need for a market to exchange foreign currency. If a U.S. automobile agency wants to import a BMW from Germany, it will need marks (the currency of Germany) to buy the BMW in Germany. Conversely, if a U.S. agribusiness sells wheat in Germany, it will receive German marks. As a result, a financial market for German marks develops inside the United States. Matching this demand and supply of marks sets the exchange rate of dollars for marks—that is, the dollar price of a mark inside the United States.

At the same time, in Germany there are firms that export to and import from the United States, and so a mark price of dollars develops inside Germany. If the two exchange markets get out of sync, international currency traders buy dollars or marks in one of the markets and transfer them to the other market. The traders will make a profit until both markets are synchronized, a process known as arbitrage.

International finance involves more than just foreign exchange transfers caused by exports and imports. Currencies also flow into and out of a country because banks make international loans, multinational corporations and mutual funds invest in foreign countries, immigrants transfer labor earnings to relatives in their home countries, nonprofit organizations such as Save the Children transfer donations, and governments build military bases or give foreign aid.

■ NORTH-NORTH CAPITAL FLOWS

While only 15 percent of the world's population live in MDCs, over 80 percent of the world's capital flows among these countries. This is due to the fact that about 80 percent of the world's production of goods and services (gross domestic product, GDP) is produced in the MDCs (World Bank 1996; Graham 1996). The history and contemporary situation of these North-North capital flows (among the MDCs) are discussed in this section. In the following sections, we examine capital flows between the MDCs and LDCs, which are much poorer and contain the vast majority of the world's population.

■ Historical Stages of the Growth of International Finance

During the nineteenth century, international finance was based on the gold standard, in which all countries agreed to exchange their national currencies for gold. The European nations sent capital to finance the development of emerging MDCs like the United States, or to their colonies to build the facilities necessary to extract mineral and agricultural products, which were then sent back to the colonizing country.

Economic competition among the MDCs was an important cause of World Wars I and II. After World War I, Russia became communist. Also, England and France demanded large war reparation payments from Germany, which depressed the German economy and contributed to the rise of Hitler. In the early 1930s, countries also began to competitively devalue their currency in relation to gold. The international economy collapsed and with it the use of the gold standard. These factors contributed to the worldwide Great Depression and to World War II.

The United States emerged from World War II as the world's dominant economic power. In the early 1950s, the total GDP of *all* the MDCs outside the United States was equal to only 40 percent of the United States' GDP. As a result of this emergence, the United States had a very strong influence over the terms of international trade and finance during the early postwar period. Other countries essentially pegged the exchange rate of their currency to the U.S. dollar.

By the late 1960s, the European countries and Japan had rebuilt their economies; most of the LDCs were no longer political colonies; the newly industrializing countries (NICs) such as Korea, Taiwan, and Brazil were emerging on their own; and an economically powerful USSR was supporting the third world liberation movements for its own political-economic reasons. All these factors challenged the economic supremacy of the dollar, and in 1973 the United States was forced to release the dollar from a fixed exchange rate, allowing its value to fluctuate according to world markets. In response, most other countries stopped fixing the exchange rate of their currency to the dollar, resulting in the present global system of flexible exchange rates.

In the 1980s, national governments accelerated the deregulation of banks and international finance. Although this deregulation led to a growth burst of international capital flows, many fear it also led to instability in the system (Korten 1996).

■ Contemporary Flows Among the MDCs

Types of flows. Currently there are four broad types of private capital flows among countries: (1) Multinational corporations carry out *foreign direct investment* (FDI). A multinational corporation is a business with at least one subsidiary or joint-ownership company located in a foreign country. Basically, FDI occurs when a multinational owns part of or makes a loan to its foreign affiliate. For example, when Honda builds a plant in Ohio, this is FDI. (2) *International loans* are money lent by commercial banks or other financial institutions, like insurance companies, to private corporations or governments in another country. For example, when the Chase Manhattan Bank of New York lends money to enable British Airways to expand its service to Russia, this is a foreign loan. (3) With *foreign portfolio investment,* investors buy stock in a foreign corporation on the stock exchange of that country. For example, when a French mutual investment fund purchases Volkswagen stock on the Frankfurt stock exchange, this is foreign portfolio investment. (4) Finally, there are *international currency* flows. These flows pay for exports and imports of goods and services and support the other types of capital flows.

Size of flows. According to a UN estimate, the total value of FDI in 1993 was $4.2 trillion. About 80 percent of this FDI was among the MDCs. There was a phenomenal growth in FDI in the late 1980s and early 1990s. The total accumulated book value of FDI tripled between 1985 and 1993; correcting for inflation, FDI doubled in just eight years. The vast majority of investment, however, remains in the home country; in 1993, FDI was still only 4.1 percent of total investment (Graham 1996). On the other hand, forty-seven of the top 100 economies in the world are multinational corporations, and by 1991, sales of the ten largest corporations exceeded the combined GDP of the world's 100 smallest countries (WFA 1996).

Global data on foreign loans is hard to find, but according to the International Monetary Fund (1991), the total value of foreign loans outstanding in 1989 was $917 billion. Seventy percent of these loans were among the MDCs. The total value of foreign loans increased 460 percent between 1983 and 1989; correcting for inflation, it tripled in just six years. International currency flows are estimated to be between $800 billion and $1 trillion *per day*. These currencies flow back and forth between countries over a year, so the annual net flows are much less than these daily figures times 365. To put the value of currency flows in perspective, the daily value of imports and exports is $20 billion to $25 billion. A substantial part of these currency flows is therefore caused by currency movements for arbitrage and speculation to earn financial profits (Korten 1996).

The principal contemporary issue for international finance is regulation of FDI and other international financial markets by national governments or by international agencies.

■ Regulation of Foreign Direct Investment

Those arguing for the free flow of FDI (little or no regulation) state that total global output increases when multinationals are free to invest in the most profitable projects and countries. They also believe that when a multinational invests in a foreign country, it brings new technology and new management and marketing styles, provides jobs, and increases that country's output. Those arguing against free flow of FDI state that these multinationals engage in cutthroat competition that harms or destroys nationally owned companies. They also believe that multinationals take profits made in the host country and return them to the home country, and that foreign multinationals have primary allegiance to their home country's government and not to the host country's. Other critics fear that multinationals do not have allegiance to, nor are they controlled by, any country.

In the late 1980s, some people in the United States complained that Japan was buying their country. Indeed, there was a rapid increase in Japanese FDI in the United States during that time. Since the 1950s, however, the U.S. multinationals have been the primary source of FDI. The difference is that, since the 1980s, other countries have also been doing a large volume of FDI, which is now occurring within the United States as well. It is also interesting to note that Britain, not Japan, is the major foreign investor in the United States.

■ Regulation of Other International Financial Markets

All agree that the international financial system has gone through some serious shocks during the post–World War II period. The first was in 1973, when the United States was forced to convert to flexible exchange rates. The second was the third world debt crisis in 1982, which is discussed below. Third, with the U.S. stock market crash in 1987, there was a flight of foreign capital from the United States, which destabilized other markets.

This destabilization was stemmed only with coordinated action by key central banks around the globe. Fourth, in 1995, Japanese land values and its stock market crashed, and the value of the Mexican peso collapsed. Both again created serious adjustment problems for the international financial system—problems that were minimized only by a series of emergency ad hoc agreements among key national government finance ministers and by the U.S. government's willingness to provide billions of dollars to support the peso to protect the investments of U.S. corporations in Mexico. Fifth, in 1997, there were dramatic sell-offs of stock—first in the Thailand stock exchange and then in the Hong Kong stock exchange. Both of these led to drops in stock values in the major international financial markets. Sixth, in 1998, there have been repercussions on international financial markets from the recent slow economic growth and dramatic foreign exchange devaluations of several of the Asian countries, including South Korea, Taiwan, Indonesia, and Thailand.

There is disagreement on how to minimize the effects of a financial collapse in one country on the global economy. Some argue that increased regulation will slow the great increase in flow of international capital that is integrating the global economy. They also argue that ad hoc agreements have solved the problems of the past and that they can do so in the future. Others argue for increasing the institutional structures for coordinating national policies. Still others say that more is needed than coordination, and that we ought to set up something like a world central bank.

These are not simple theoretical questions. In the early 1980s, social democratic governments in France and Spain were pursuing a macroeconomic policy of low unemployment with slightly higher inflation. The United States and Britain, however, were willing to tolerate higher unemployment rates, partly to keep wages lower, and they preferred lower inflation rates. Capital movements out of France and Spain to the United States and Britain contributed to the failure of the French and Spanish policies. Those arguing for regulation to prevent personal abuse cite the example of the 1995 scandal surrounding a twenty-eight-year-old investor in Singapore. This young man sitting at a computer terminal made millions of dollars in profits for his bank and salary and bonuses of several million dollars for himself. Then, over roughly a four-week period, he bet $29 billion of the bank's money and ran up losses of $1.3 billion. As a result, the venerable 223-year-old Barings Bank of England went bankrupt, and the Tokyo stock exchange index dropped (Korten 1996). Those arguing for deregulation say these results are as they should be; if people, organizations, or even countries behave irresponsibly, they should pay for it.

These potential problems leave us with a series of questions that are not easily answered. Should we regulate? If so, how? Can a balance be struck between international regulation and national autonomy? If we use regulation, will it be controlled by large corporations and special interests for their own benefit, or can regulation be done to benefit all? Many, including this author, believe markets need to be open but responsible. To ensure responsibility some form of democratically controlled regulation is desirable. But that is not easy to achieve. For more on the debate about regulation see the suggested readings at the end of this chapter (especially Graham, Korten, and WFA).

■ NORTH-SOUTH CAPITAL FLOWS

Having examined North-North capital flows—that is, flows among the MDCs—we now turn our attention to North-South capital flows between the 15 percent of the world's population in the MDCs and the 85 percent of the world's population in the LDCs. The most important areas of controversy for the LDC capital flows are (1) foreign direct investment, (2) the third world debt problem (LDC problems in repaying previous foreign loans), and (3) foreign aid.

■ Foreign Direct Investment

Former president Mobutu of Zaire (now Democratic Republic of Congo) referred to his country as "underequipped" rather than "underdeveloped." This casual remark reflects a long-standing debate among economists as to whether foreign direct investments from MDCs to LDCs are a cause of development, simply an ingredient of it, or an element that retards development. Some agree with Mobutu, who argues that flows from MDCs into LDCs are the missing element. They also believe that international financial institutions, such as the World Bank and regional development banks, are set up to provide LDCs with capital when and where private investors fail to do so. Others argue that it is the institutional deficiencies in LDCs themselves that make these countries less than ready to receive capital from abroad. In Zaire, for example, Mobutu, by looting foreign companies to sustain his own power and wealth, caused these companies to abandon their factories in Zaire and refuse to return.

The primary source of any country's capital is actually not foreign capital flows but domestic saving. In order for investment to promote development, people must save. If the savings of a given country are not enough to finance the needed investment, it is possible to draw on other countries' savings through their foreign investment.

Historically, investment has flowed from less to more profitable countries. European capital flowed to the United States in the nineteenth century because the frontier and expanding population made such investment profitable. The investment climate, however, must be receptive. If President

Mobutu stole the assets of foreign companies or prevented them from realizing profits, foreigners likely would stop investing in Zaire. In contrast, in the "four dragons" of East Asia (South Korea, Taiwan, Hong Kong, and Singapore), the investment climate has been made attractive, helping economic development to surge in these countries in the past thirty years.

Many in the LDCs say there is good reason they have not made an attractive investment climate. Multinational corporations (MNCs) extract raw materials and profits from their host countries and bribe government officials for special privileges. Thus, they argue, foreign investment detracts from economic development rather than promotes it. While no one has a definitive answer, this chapter tries to shed light on the question.

Size of the flows. In the early 1990s, private investment from MDCs into LDCs suddenly surged. Table 7.1 shows the aggregate net long-term flows of investment capital into LDCs from 1989 to 1994. Official capital denotes capital from governments and international agencies. Note that, while official grants and loans have remained approximately the same during these six years, private investment has increased more than fourfold. This surge in private capital flows has been concentrated in about twenty countries, most of which are middle-income countries in East Asia and Latin America, and also in two large low-income countries, China and India (World Bank 1995). The increase in private flows has been spread among all the categories: debt (or private lending), direct investment, and portfolio investment (purchase of stocks and bonds of various corporations). With the 1995 collapse of the Mexican stock market, however, portfolio investment in the LDCs was sharply curtailed.

Policy issues. To understand why this surge in private flows occurred, and to guess whether it will happen again, one must look at the conditions of the less developed world two decades ago. At that time, LDCs held a sharp antagonism toward foreign investment. LDC governments restricted the ability of multinationals to export, import, repatriate capital, set prices, and negotiate with labor unions. They required permission for ordinary transactions that could be freely undertaken in MDCs; and bribery was often the only way to obtain such permissions. Many of these restrictions applied to local as well as foreign firms, resulting in much business being conducted under the table.

These restrictions on investment, both domestic and foreign, were accompanied by intense antagonism toward multinational corporations, much of which was justified. In 1954, for example, the U.S. government gave military support to Guatemalan rightist forces who overthrew the reformist president, Jacobo Arbenz. He had planned land reform that was to confiscate and redistribute some of the land holdings of, among others, the United Fruit Company. United Fruit was a U.S.-owned corporation that

Table 7.1   Aggregate Net Long-Term Resource Flows to LDCs, 1989–1994
(billions of U.S.$)

	1989	1990	1991	1992	1993	1994
Official						
Grants	19.2	28.7	32.6	29.9	30.1	30.5
Loans (net)	23.4	29.2	29.2	20.4	23.8	24.0
Total official	42.6	57.9	61.9	50.3	53.9	54.5
Private						
Debt (net)	12.7	15.0	18.5	42.4	45.7	55.5
Direct investment	25.7	26.7	36.8	47.1	66.6	77.9
Portfolio securities	3.5	3.8	7.6	14.2	46.9	39.5
Total private	41.9	45.5	62.9	102.7	159.2	172.9
Total official and private	84.5	103.4	124.7	153.0	213.2	227.4

Source: World Bank, World Debt Tables (Washington, DC: World Bank, 1995).

had close ties to the Dulles brothers, who were leaders in the U.S. State Department and Central Intelligence Agency (CIA). As another example of justified antagonism, when Salvador Allende was elected president of Chile in 1970, U.S. multinationals feared that his government would expropriate their investments in Chile. In 1972, Allende was murdered and the new government, a military dictatorship, was quickly endorsed by the U.S. government. This dictatorship was friendly to foreign business, and it was supported by the United States for almost twenty years.

Advocates for the MNCs argue that most MNCs pay their workers well, negotiate fairly with unions, do not cheat on taxes, and refrain from bribery. One investigation supporting this view is an International Labour Organization study that reported that MNCs pay wages, on average, twice as high as local businesses and offer significantly more fringe benefits: housing, education, hospitalization, and health services (ILO 1976).

By the mid-1980s, the LDCs began developing a more favorable climate for FDI. Not only was the Soviet Union failing, but many LDCs were deeply in debt, corruption-ridden, and had state industries unable to face world competition. Also, international agencies were willing to assist only if the LDC governments would sell off unproductive state enterprises, free up exchange restrictions and prices, and open their doors to foreign private enterprise. These have been important causes of the recent foreign private investment increases in LDCs. Nevertheless, a problem remains: virtually all this investment is concentrated in middle-income countries, with almost none going to the very poor ones.

■ The Debt Problem

Causes and size. When the Organization of Petroleum Exporting Countries (OPEC) raised the price of oil in 1973, the impact was felt primarily in the then non-oil-producing less developed countries (sometimes known as NOPEC). Their net payments for oil purchased from OPEC increased by $11.4 billion between 1970 and 1975 (Powelson 1977). While industrialized countries increased their net oil payments by much more ($62.8 billion) in the same period, they did not feel the same monetary or balance of payments effects, since OPEC mostly invested their windfall proceeds back into those more developed countries and deposited reserves in their banks. Relative to their GDPs and balance of payments, the impact of increased oil prices was therefore most severe on the NOPECs.

To sustain the NOPEC development programs and to finance their balance of payments deficits, both governments and commercial banks in industrialized countries and international agencies made special loans to NOPEC governments and central banks. Since nothing had happened to increase NOPEC exports or to attract increased investment, their indebtedness piled up, marking the beginnings of the modern debt crisis.

Strangely, the debt crisis spread to oil-producing countries as well. On the strength of further revenue prospects for the indefinite future, these countries increased their borrowing for "development projects." Mexico, Nigeria, and Indonesia are particularly cited. When oil prices tapered off in the 1980s, these countries were unable to repay their debts. Since no great increase in development was realized in any of these three countries, it has been supposed that the borrowings mainly financed extravagant and wasteful projects or lined the pockets of political and economic elites.

Between 1989 and 1994, developing countries' outstanding foreign debt increased from $1,369 billion to $1,945 billion, equaling about 7 percent of their gross domestic product (World Bank 1995). In low-income areas, this constitutes an enormous repayment burden; the interest charges alone subtract from the amounts available for domestic consumption and investment.

Debt restructuring. For reasons that combine political interests, compassion, and the need for world stability, industrialized nations, the World Bank, the International Monetary Fund (IMF), and other international agencies have been seeking ways to reschedule and reduce third world debt. In the early 1980s, some key banks, such as the Bank of America, held a substantial percentage of their loan portfolio in loans to LDCs. When many countries found that they could not repay their loans, these large banks in the MDCs were forced to write off part of this debt as bad loans. These write-offs created a global financial crisis that contributed to the recessions of the early 1980s in both the United States and Europe.

Third world debt has continued to increase over time. One cause contributing to this increase has been the high interest rates in the MDCs, particularly in the 1980s. For a loan with a compound interest rate of 10 percent, for example, the debt doubles in about seven years, even if no new loans are taken out. The indebted countries, for their part, have been eager to borrow more to sustain their domestic consumption and investment, to alleviate balance of payments deficits, and to repay old loans.

Given these problems, the international financial community has sought ways to restructure the economies of indebted countries in order to contain increases in indebtedness. As a condition for its assistance to indebted LDCs, the IMF demanded that these countries reduce market regulations and have realistic exchange rates. As time went on, privatizing inefficient state enterprises and balancing government budgets were added to the list. Such conditions for refinancing are called conditionality.

Conditionality issues.   IMF conditionality, also required by the World Bank and governments of MDCs, is very controversial. Many view these conditions as necessary for the increased economic growth needed by LDCs to repay their debts. Others, however, hold that conditionality is big banks from rich MDCs interfering in the internal politics of LDCs.

In the LDCs, the public complains that the poor suffer from the resulting austerity, since unemployment increases and government services are reduced; at time same time, internal bank loans are cut and budgets are balanced to reduce inflation. Public protests, demonstrations, and riots have resulted. The IMF argues that the suffering is inevitable due to wasteful government policies, and that the IMF program is actually diminishing suffering by smoothing over the necessary adjustments. More sage economists point out that elites in the developing countries have squandered the borrowed funds and the poor are now suffering from the austerity needed to bring order into government accounts.

Despite the many debt forgiveness plans proposed by governments and international lending agencies, the debt of developing countries continues to increase, creating a world problem of crisis proportions. Indeed, net total private and official flows from the MDCs to the LDCs was negative from 1983 to 1991 (UNDP 1994). In other words, the poor were sending more capital to the rich than they were receiving from them. Twenty countries have reached such a high debt level that they are now excluded from international capital markets (*New York Times* 1996). Debt is relatively most severe in African countries, where the standard of living and degree of economic development are among the lowest in the world.

■ Foreign Aid

Seventy percent of foreign aid is bilateral (country to country); the remainder is multilateral (countries give aid via international agencies such as the World Bank and UNESCO, the United Nations Educational, Scientific and Cultural Organization) (UNDP 1994). About 10 percent of foreign aid comes as grants and does not require repayment; but most comes as concessionary loans, with lower interest rates and longer repayment periods than private sector loans. Nongovernmental organizations, such as Save the Children and the International Red Cross, also give aid to LDCs. Such aid is about 10 percent as large as official development assistance (ODA, bilateral plus multilateral). Finally, there is disguised foreign aid such as special permits for an LDC to export to an MDC. For example, Russia supported Cuba for years by buying sugar from them at higher than world market prices.

Size of foreign aid flows. Table 7.2 shows ODA by the principal donor countries. The Development Assistance Committee (DAC) is an organization of the twenty-two largest MDCs that give foreign aid. Among the DAC countries, the United States gives the second largest total amount of ODA, but it gives the *lowest* percentage of its gross national product (GNP). Among these countries, the United States has the largest decrease in ODA as a percentage of GNP between 1965 and 1991.

Total ODA in 1994 was $59.2 billion, an average of 0.30 percent of DAC countries' GNPs. In the 1970s, there was an international campaign to increase foreign aid to 1.0 percent of the GNP of the DAC countries, but from Table 7.2 we see that on average it decreased by 0.13 percent of GNP between 1965 and 1991. The United Nations Development Programme (UNDP) estimates that with about $40 billion more of ODA directed to basic human needs per year over the next ten years, the globe could provide universal access to safe drinking water, primary health care, basic education, and a basic family planning package to all willing couples (UNDP 1994).

Table 7.2 Official Development Assistance by Principal Donor Countries

Country	% of GNP (1994 U.S.$)	Billions of U.S.$ 1994	% of GNP 1991 Minus % of GNP 1965
Sweden	0.96	1.8	+0.71
France	0.64	8.5	−0.14
Canada	0.43	2.4	+0.26
Germany	0.34	6.8	+0.01
United Kingdom	0.31	3.2	−0.15
Japan	0.29	13.2	+0.05
United States	0.15	9.9	−0.38
22 DAC country average (total)	0.30	(59.2)	−0.13

Sources: UNDP, *Human Development Report* (New York: Oxford University Press, 1996); World Bank, *Investing in Infrastructure: World Development Report 1994* (New York: Oxford University Press, 1994).

Table 7.3   Official Development Assistance by Recipient Regions, 1994

Region	Per Capita U.S.$	Billions of U.S.$	% of Recipient's GDP
Arab states	29	6.9	3.5
Sub-Saharan Africa	32	18.9	10.5
East Europe and central Asia	24	1.9	1.0
Latin America and Caribbean	13	5.8	0.4
South Asia	6	7.2	1.5
Southeast Asia and Pacific	12	5.5	1.3
Average (Total)	12.1[a]	(59.2)	1.3[a]

Source: UNDP, *Human Development Report* (New York: Oxford University Press, 1996).
Note: a. 1991 data from World Bank, *Investing in Infrastructure: World Development Report 1994* (New York: Oxford University Press, 1994).

Table 7.3 shows ODA by recipient regions. The sub-Saharan African countries receive the most per capita *and* total dollars of foreign aid, and it is the highest percentage of their GDPs. In the early 1990s, private capital flows from the MDCs to the LDCs were five times larger than ODA. Just twelve countries, however, receive 90 percent of these private flows, and the forty-eight poorest countries receive only 0.4 percent of private flows. In comparison, while total ODA in 1991 was $49.4 billion, remittances back to the home country from immigrants working in MDCs was $20 billion (UNDP 1994).

Arguments for giving and receiving foreign aid. A political reason *for giving aid* would be so that an MDC could support a particular LDC government that has policies favorable to the granting MDC. A military reason would be to economically support military allies in a region of conflict, such as the United States giving aid to Israel. (Note that the ODA figures in the preceding section and tables do not include direct military aid.) An economic reason would be to open markets in the LDCs for the donor country's businesses. Humanitarian reasons would be to prevent malnutrition and diseases and to provide basic needs for all. A larger humanitarian reason would be to redistribute resources from the prosperous MDCs to the poor LDCs in order to alleviate the maldistribution of global output, thereby encouraging peace and democracy throughout the globe.

Data from a United Nations report helps us to understand the relative strength of these various motives for giving aid (UNDP 1994). In 1993, recipient countries that had military spending greater than 4 percent of GDP received $83 per capita, while those with military expenditures below 2 percent of GDP received $32 per capita. Also, $280 and $176 per person in poverty went to Egypt and Israel, respectively, both of which are quite prosperous but are strategic allies of the United States; $19 per poor person went to Bangladesh, one of the poorest countries on the globe. Seven percent of total bilateral ODA goes directly for basic needs, as does 16 percent of multilateral ODA.

An important economic justification *for receiving aid* is to acquire the capital needed for economic development. If outside funds are invested wisely, they will create jobs and output. This new income can then be used to repay the loans. Indeed, according to this view, once a country becomes sufficiently developed it becomes a capital exporter.

Arguments against giving and receiving foreign aid. Some arguments *against giving aid* are: (1) It is inappropriate to interfere in another country's internal affairs. (2) Countries need to get their own houses in order before seeking aid, otherwise the aid is wasted or, worse, it fosters internal corruption. (3) Money would be better spent on the poor at home—why send it abroad? If all foreign aid were spent on domestic social welfare expenditures in the MDCs, these expenditures would increase from 15 percent of their GDPs to 15.3 percent (UNDP 1994).

Some arguments *against receiving aid* are: (1) Countries should have the right to choose their own style of economic development, and foreign aid skews these choices to the visions of the grantor. The vast majority of all of foreign aid, for example, goes for large-scale modernization projects. (2) Foreign aid creates enclaves of international mining, agribusiness, and unskilled manufacturing. These enclaves have little positive influence in helping modernize the domestic economy. Rather, they exist to export national resources to the MDCs. (3) Much foreign aid is "tied" aid, meaning that the aid has to be spent in the granting MDC. (4) Most aid is given as loans, creating future indebtedness for a country. (5) Some argue that even humanitarian aid given at times of famines results in only a temporary fix. In the long run, such aid harms the internal adjustment process necessary for a long-term solution.

Some in the LDCs argue "trade not aid." They claim that protectionism in the MDCs deprives them of export markets important for their development (see Chapter 6). With 85 percent of the world's population, the LDCs had only about 24 percent of world exports in 1994. As with foreign direct investment, this trade was concentrated in the more prosperous LDCs.

■ CONCLUSION

Along with modern technology, we humans have invented our international financial institutions, and like our use of technology, we can use these economic institutions for good or for ill. International capital flows facilitate the worldwide movement of goods, services, and savings. These flows can facilitate global economic growth, integration, and justice. If, however, we do not stay in contact with some higher principles (which some would call our Inner Light, or God) these flows can be used to gain economic power over others and to create instability and injustice in the international economy. Some argue that global governance is necessary to ensure a stable and just flow of international capital throughout the globe, while others argue that these goals will be better served by financial markets free of regulation.

■ QUESTIONS

1. Discuss how the global financial system is an institution created by humans, which we can use for good or for ill.

2. Do the advantages of foreign multinational investment on U.S. soil outweigh the disadvantages?

3. On balance, does private foreign direct investment promote or discourage economic development in the LDCs? Discuss strategies that LDCs might adopt to make FDI fit their development aspirations better without severely harming incentives for FDI.

4. What is conditionality? Is it necessary, or is it oppressive to the LDCs? How would you solve the LDC debt problem, keeping in mind stability of the international financial system as one of your goals?

5. Which do you think is more desirable: bilateral or multilateral foreign assistance?

6. Should the United States give more or less foreign aid? If more aid, what are appropriate motives for giving aid, and where and for what should the aid go?

7. What are the advantages and disadvantages of increasing the extent of global governance over the international economy?

■ SUGGESTED READINGS

Balaam, David, and Michael Veseth (1996) *Introduction to International Political Economy.* Upper Saddle River, NJ: Prentice Hall.
———— (1996) *Readings in International Political Economy.* Upper Saddle River, NJ: Prentice Hall.
Epstein, Gerald, Julie Graham, and Jessica Nembhard, eds. (1993) *Creating a New World Economy: Forces of Change and Plans for Action.* Philadelphia: Temple University Press.

Graham, Edward M. (1996) *Global Corporations and National Governments.* Washington DC: Institute for International Economics.

Korten, David C. (1996) *When Corporations Rule the World.* West Hartford, CT: Kumarian Press.

Todaro, Michael P. (1997) *Economic Development.* New York: Addison-Wesley.

United Nations Development Programme (1994) *Human Development Report.* New York: Oxford University Press (see chapter 4: "A New Design for Development Cooperation").

WFA (World Federalist Association) (1996) *The Global Economy, Part 2: TNCs and Global Governance.* Washington, DC: WFA.

8

Poverty in a
Global Economy

Don Reeves

- Poverty is a mother's milk drying up for lack of food, or kids too hungry to pay attention in school.
- Poverty is to live crowded under a piece of plastic in Calcutta, or huddled in a cardboard house during a rainstorm in São Paulo, or homeless in Washington, D.C.
- Poverty is watching your child die for lack of a vaccination that would cost a few pennies, or never having seen a doctor.
- Poverty is a job application you can't read, or a poor teacher in a run-down school, or no school at all.
- Poverty is hawking cigarettes one at a time on jeepneys in Manila, or being locked for long hours inside a garment factory near Dhaka or in Los Angeles, or working long hours as needed in someone else's field.
- Poverty is to feel powerless—without dignity or hope.

■ DIMENSIONS OF POVERTY

Poverty has many dimensions. Religious ascetics may choose to be poor as part of their spiritual discipline. Persons with great wealth may ignore the needs of those around them, or may miss the richness and beauty of nature or great art and remain poor in spirit. But this chapter is about poverty as

This chapter is adapted from an earlier work by the author in *Hunger 1995: Causes of Hunger* (BFW Institute 1994) and is used with permission of BFW Institute.

the involuntary lack of sufficient resources to provide or exchange for basic necessities—food, shelter, health care, clothing, education, opportunities to work and to develop the human spirit.

Globally, poor people disproportionately live in Africa. The largest number live in Asia. A significant number are in Latin American and Caribbean countries. Up to 80 percent of the people in several sub-Saharan Africa countries and Haiti are poor, and nearly one-third of all peoples in developing countries together are poor. (The situation worldwide is shown in Figure 8.1.)

But no place on the globe is immune to poverty. The United States, some European countries, and Australia also have large blocs of poor people. With few exceptions, the incidence of poverty is higher in rural than in urban areas but is shifting toward the latter. Nearly everywhere, women and girls suffer from poverty more than do men and boys; infants, young children, and elderly people are particularly vulnerable. Cultural and discriminatory causes of hunger are immense; the difficulties in changing long habits and practices should not to be underestimated (BFW Institute 1994).

In this chapter, we look first at ways in which poverty is measured. Then we look at approaches to reducing poverty in the context of a global economy, especially the relationship between economic growth and inequality. Finally, we examine a series of policy choices developing country societies might consider as they attempt to reduce poverty.

Figure 8.1   Number and Percentage of Poor People Worldwide, 1987–1993

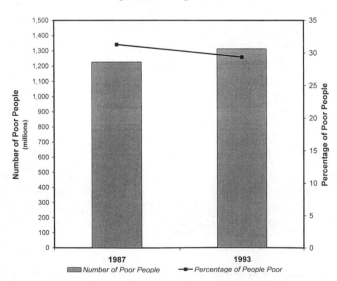

Source: World Bank, *World Development Indicators* (New York: Oxford, 1997).

■ MEASURING POVERTY AND INEQUALITY

Poverty is not the same in the United States, or Poland, or Zimbabwe. It will often be described differently by supporters or critics of a particular regime. Poverty does not lend itself to an exact or universal definition. Deciding who is poor depends on who is measuring, and where, and why. So, first, let us look at how poverty and inequality are measured.

Poverty is usually measured by income or consumption. By the most widely used measure, the World Bank estimates that worldwide 1.3 billion people live on incomes equivalent to less than U.S.$1 per day (World Bank 1997a). The majority of these people chronically lack some or all basic necessities. The rest live so close to the edge that any emergency—illness, work layoff, drought—pushes them from just getting by into desperation.

In the United States, poverty is defined as three times the value of a thrifty food plan devised by the U.S. Department of Agriculture, adjusted for family size—$16,036 for a family of four in 1997. Some critics say it is higher than necessary, partly because certain government program benefits—such as Medicare, housing subsidies, and school meals—are not counted. But poor people themselves feel hard pressed. The thrifty food plan was worked out in the early 1960s to address short-term emergencies. Although it is adjusted annually for changes in food prices, other costs, particularly housing, have grown faster than food costs since the plan's base year (1955); so the threshold has represented a gradually declining standard of living.

The selection of the poverty threshold often makes a dramatic difference in the observed poverty rate. The World Bank, based on its cutoff of $1 per person per day, estimates that in 1993, 350 million of China's total 1.2 billion people were poor. The Chinese government, using a lower cutoff point, claims that only about 100 million of its people were poor (World Bank 1997b).

Two other widely used income measures are per capita gross domestic product (GDP) and gross national product (GNP). GDP is the value of all goods and services produced within an economy; GNP equals GDP plus or minus transfers in and out of the economy, such as profit paid to foreign investors or money sent home by citizens working abroad.

Among countries with populations exceeding 1 million, the World Bank counts fifty-eight low-income economies with an annual per capita GNP from $80 to $765 (1995). At the other end are twenty-five high-income economies with per capita GNP from $9,386 to $41,210. In between are sixty-seven middle-income economies with per capita incomes between $766 and $9,385 (World Bank 1997a).

GNP (or GDP) provides a quick measure of the capacity of an economy overall to meet people's needs. It also represents the pool from which savings and public expenditures can be drawn. But GNP is seriously flawed as a measure of poverty or well-being because it gives no information about the quality of the production or the distribution of income within the country.

First, GNP/GDP fails to distinguish among types of economic activity. Manufacturing cigarettes, making bombs, and running prisons are scored as contributing to GNP/GDP the same as making autos, teaching school, building homes, or conducting scientific research. Second, many goods and services generate costs that are not reflected in their prices—polluted air from manufacturing or illness from overconsumption, for example. Third, many nurturing and creative activities—parenting, homemaking, gardening, and home food preparation—are not included because they are not bought and sold. At best, GNP and GDP figures include only *estimates* for food or other goods consumed by producers, unpaid family labor, and a wide range of other economic activities lumped together as the informal sector. Illegal or criminal activities, such as drug-dealing or prostitution, are generally not included in estimates but nonetheless contribute to some people's livelihood.

◼ Purchasing Power Parity

GNP/GDP figures for various countries are usually compared on a currency exchange basis. The per capita GNP in Bangladesh, at 8,800 taka, could be exchanged for U.S.$220.

But 8,800 taka will buy more in Bangladesh than $220 will buy in the United States, primarily because wages are much lower. Thus, the World Bank and the United Nations Development Programme (UNDP) have adopted a new measure—purchasing power parity (PPP)—which estimates the number of dollars required to purchase comparable goods in different countries. Bangladeshi PPP is estimated at $1,330, rather than $220 (UNDP 1997).

PPP estimates make country-to-country comparisons more accurate and realistic and somewhat narrow the apparent gap between wealthy and poor countries. Even so, vast disparities remain. PPPs of $21,000 to $27,000 per capita—as in the United States, Switzerland, and Canada—are forty to fifty times those of Ethiopia and Tanzania, at $450 and $640 (World Bank 1997a).

◼ Inequality

Estimates of poverty and well-being based on estimated GDP are at best crude measures. GNP, GDP, and PPP are all measured as country averages. But because poverty is experienced at the household and individual level, the distribution of national incomes is a crucial consideration.

Detailed and accurate information is necessary for targeting an-tipoverty efforts and particularly for assessing the consequences of policy decisions in a timely fashion. But census data as comprehensive as that for the United States are a distant dream for most poor countries. Many of them do not keep such basic records as birth registrations and may have only a guess as to the number of their citizens, let alone details about their conditions. Representative household surveys are the only viable tool for most countries for the foreseeable future.

Household surveys, to be useful—especially for comparison pur-poses—need to be carefully designed, accurately interpreted, and usable for measuring comparable factors in different times, places, and circum-stances. Private agencies, many governments, and even some international agencies are tempted to shape or interpret surveys to put themselves in the best light. Users of survey results need to be keenly aware of who con-ducted the survey and for what reasons.

Globally, we have accepted gross income inequality. The most used measure of inequality compares the income of the richest one-fifth, or quintile, of each population with that of the lowest quintile. The wealthiest one-fifth of the world's people control about 85 percent of global income. The remaining 80 percent of people share 15 percent of the world's in-come. The poorest one-fifth, more than a billion people, receive only about 1.4 percent. The ratio between the average incomes of the top fifth and the bottom fifth of humanity is 60 to 1 (see Figure 8.2).

Using recently revised U.S. census data, in 1980 the richest one-fifth of U.S. households received 44.1 percent of the total income, while the poorest one-fifth had 4.2 percent—a ratio of 10.5. By 1994, the ratio had widened; the top one-fifth had 49.1 percent, the bottom one-fifth only 3.6 percent, and the ratio had widened to 13.6.

Using older World Bank data, the ratio between the rich and poor quintile's incomes in the United States stood at 8.9 in 1985. Among other

Figure 8.2 Distribution of World Income

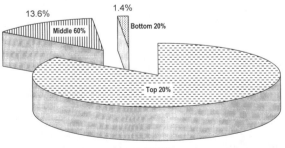

Source: United Nations Development Programme, *Human Development Report* (New York: Oxford University Press, 1997).

industrial nations, the ratio ranged from 4.3 in Japan up to 9.6 for Australia and the United Kingdom. Among the low-income countries with estimates available, the ratios ranged from 4.0 for Bangladesh and 5.0 for India up to 32.1 for Brazil (World Bank 1997a).

Differences in income distribution make a big difference to poor people. Thailand's per capita GDP is only slightly larger than Brazil's, but the poorest 20 percent of the population in Thailand has more than three times as much purchasing power. In Indonesia, with less than 60 percent of Brazil's per capita GDP, poor people have more than twice the purchasing power. Even in Bangladesh, with per capita GDP less than one-fourth that of Brazil, poor people are estimated to have as much purchasing power as in Brazil (see Table 8.1).

■ Direct Measures of Well-Being

Other indicators measure well-being even more directly than income or poverty rates: for example, infant or under-five mortality rates, life expectancy, educational achievement, and food intake.

Hunger. The Food and Agriculture Organization of the United Nations (FAO) attempts to measure and estimate shortfalls in food consumption. In their *Sixth World Food Survey,* released in late 1996, the FAO estimates that the absolute number of people in developing countries who consumed too little food declined slightly over the past two decades—from about 918 million in 1969–1971 to about 841 million in 1990–1992. Because population increased rapidly over the period, however, the proportion of people hungry in developing countries declined from about 35 percent to about 20 percent.

The most dramatic gains in reducing hunger over the period were in East and Southeast Asia, where the percentage of hungry people dropped from 41 to 16 percent and the number by nearly half—from 476 million to 269 million. Less dramatic gains by both measures were recorded in the Middle East and North Africa. The proportion declined, but the absolute

Table 8.1   Poverty Impact of Income Distribution, Selected Countries, 1994

Country	GNP PPP$/Capita 1994	GNP–PPP$ 1994 Lowest 20%	Highest 20%	Ratio Highest 20%/ Lowest 20%
Thailand	6,970	1,951	18,366	9.4
Brazil	5,400	567	18,225	32.1
Indonesia	3,600	1,566	7,326	4.7
Bangladesh	1,330	625	2,520	4.0

Source: World Bank, *From Plan to Market: World Development Report 1996* (New York: Oxford University Press, 1996).

number increased slightly over the period in South Asia and Latin America and the Caribbean. In Africa, the percentage of hungry people increased from 38 to 43 percent, while the number soared from 103 to 215 million (see Figure 8.3) (FAO 1996).

Figure 8.3   Distribution of Undernourished People by Developing Region, 1969–1971 and 1990–1992

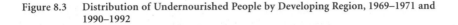

1969-71: 918 million undernourished

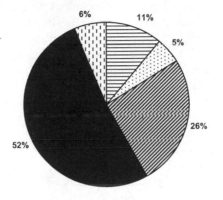

1990-92: 841 million undernourished

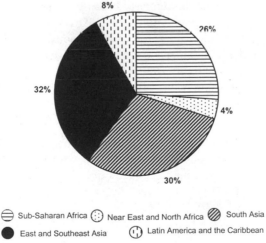

Source: Food and Agriculture Organization of the United Nations, *The Sixth World Food Survey* (Rome: FAO, 1996). Reprinted with permission.

Human development index. The United Nations Development Programme has developed a Human Development Index (HDI), which gives equal weight to three factors: life expectancy at birth, educational attainment (based on the adult literacy rate and mean years of schooling), and per capita purchasing power (UNDP 1997).

People's lives can be improved if even limited resources are focused on nutritional programs, public health, and basic education. Both China and Sri Lanka, for example, have invested relatively heavily in education and health care since independence. They rank with many industrial countries in life expectancy and educational attainment.

Most of the formerly communist countries invested in education and health care. Some former colonies continued to build on the educational systems established during the colonial era: Vietnam, Laos, Cambodia, and Madagascar (colonized by France); Guyana, Tanzania, Uganda, and Burma (colonized by Britain); and the Philippines (colonized by Spain and the United States). Several Latin American countries have emphasized education more recently: Chile, Costa Rica, Colombia, and Uruguay. In each instance, they rank higher on the HDI scale than other nations with a comparable per capita gross domestic product.

But sustaining such improvements requires steady or improving economic performance. Many of these nations have suffered recent economic downturns or are in the midst of drastic political and economic change. In the short term, at least, they are hard pressed to maintain their education and health programs.

Other nations rank much lower on the HDI scale than on a per capita GDP scale. The Middle Eastern oil-rich nations rank low in both longevity and educational attainment, particularly because of the status of women in their societies. Several African nations have extremely low educational attainment and longevity indicators, for varied reasons. Angola and Namibia have been engulfed in long independence struggles and civil war. Botswana and Gabon, although relatively rich in natural resource income, have not devoted proportional resources to education and health care services.

*   *   *

After this extensive digression to explore how poverty is measured, we turn now to the questions of why poverty is worse in some places than others and what might be done to reduce or eliminate it.

■ ECONOMIC GROWTH AND POVERTY REDUCTION IN A GLOBAL, KNOWLEDGE-BASED, MARKET ECONOMY

Individual, community, and national efforts to reduce poverty must be set in the threefold revolution during the past two decades that has transformed national markets into a truly global economy:

- The evolution of a single worldwide system of producing and exchanging money, goods, and services
- The shift from a resource-based economy to a knowledge-based economy
- The acceptance of market-based economics as conventional wisdom by most political leaders throughout the world

The nature of and possible gains and risks from this revolution are much more fully developed in Chapters 6 and 7.

■ The Global Workforce: Need for 2 Billion Jobs

The route out of poverty for most people is through new economic opportunities—jobs or business ventures. As the world's population grows one-third, from approximately 6.0 billion to 8.0 billion or more by the year 2025, the global labor force will grow even faster, by about half, from 2.7 billion to more than 4 billion workers. In addition, the International Labour Organization (ILO 1996) estimates that more than a billion workers are now unemployed or underemployed. Half the new workers have already been born, and the number of unemployed is growing annually. The pressing need, therefore, is to create 2 billion new economic opportunities during the next thirty years. Most of the new jobs or businesses will be needed in developing countries, where 95 percent of the increase in population and labor force is taking place.

Virtually all the added jobs will need to be nonfarm. Governments in developing countries, or markets, may increase incentives for food production, but farmers are likely to adopt technologies that increase their productivity and reduce farm employment even faster. More and more farmers, or their children, will seek nonfarm employment. Whether such nonfarm employment is urban or rural will depend on policy choices. Improvements in education, health care, and public infrastructure can provide some public service jobs. But most new income-earning opportunities, if they come to pass, will be in the private sector.

Every new opportunity, whether public or private, for employee or self-employment requires savings and investment—in human resources and in creating each job or business opportunity. The rate of savings and their allocation are crucial factors in determining whether enough decent income-earning opportunities can be created; these factors are determined in large measure by public policies.

■ Economic Growth and Poverty Reduction

Economic growth is often held up as the primary goal for economic development and as the means to increased employment opportunities. Some analysts, bankers, and political leaders almost equate "development" with economic growth. Most of these people expect poverty and other social problems to shrink as economies grow—the "rising tide" of economic growth will "lift all boats." Economic growth is a necessary, but not sufficient, condition for reducing poverty. The distribution of the added income is also critical. Poverty has fallen rapidly in some fast-growing economies (Korea, Indonesia, China), while not changing much in others (Brazil, South Africa, Oman).

Because poverty is experienced in households and by individuals, detailed and accurate information at that level is critical. The World Bank recently commissioned a review of 109 household surveys, covering sixty-four time comparisons between 1987 and 1993 in forty-six developing economies (Ravallion and Chen 1997). The surveys showed the following:

- Poverty rates have consistently fallen as average incomes have risen, and risen as average incomes have fallen.
- Poverty rates have not declined anywhere in the absence of economic growth.
- In developing countries, inequality increased as often as it decreased as average incomes rose.
- In transition economies (former communist states in Eastern Europe and Central Asia), during 1987–1993, inequality consistently increased at the same time as average incomes fell (see Figures 8.4 and 8.5).

Other recent studies of eight East and Southeast Asian countries show that it is possible to have both economic growth and decreasing inequality if the right policies are in place. In South Korea, for example, where per capita income has grown rapidly, the most affluent fifth of the population has about six times as much income as the poorest fifth. The ratio has narrowed slightly over the past two decades; poor people have shared in the rapid growth.

In sharp contrast, Brazil's per capita GNP was twice Korea's in 1970. Since then, its economy has grown about half as fast; by 1990, Korea's per capita GNP was twice Brazil's. But the income ratio between Brazil's poorest and richest fifth is more than thirtyfold and has increased. Poor Brazilians have scarcely benefited from growth and remain mired in deep poverty.

The Asian countries reduced, or at least did not increase, economic inequality by giving poor people the incentive and the means to improve their own earning power; examples are land reform and support for small farmers in Korea and Taiwan; high school education, especially for women, in Singapore; and manufacturing for export that raised the demand

Figure 8.4   Economic Growth and Poverty Rates, Developing and Transition Countries, 1987–1993

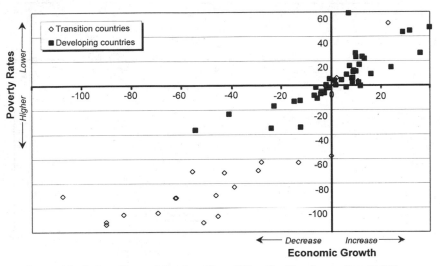

Source: Martin Ravallion and Shaohua Chen, "What Can New Survey Data Tell Us About Recent Changes in Distribution and Poverty?" *World Bank Economic Review* (Washington, DC: World Bank 1997).

Notes: Poverty rates = change in % of people poor x –100; Economic growth = log of change in average household consumption x 100.

for unskilled factory workers, plus a massive affirmative action program for the poorer ethnic groups, in Malaysia.

Declining inequality and economic growth support each other in three ways:

- As poor families' income increases, they invest more in "human capital"—more education and better health care for their own children.
- Improved health and better education, which usually accompany decreased inequality, increase the productivity of poorer workers and their communities and nations.
- Greater equality contributes to political stability, which is essential for continued economic progress.

Recent research shows that relative equality in distribution of national incomes increases the likelihood that economic growth can be sustained. Widespread participation in political as well as economic activity reduces the likelihood of enacting bad policies and permits their earlier correction (Birdsall, Pinckney, and Sabot 1996).

Figure 8.5   Economic Growth and Inequality, Developing and Transition Countries, 1987–1993

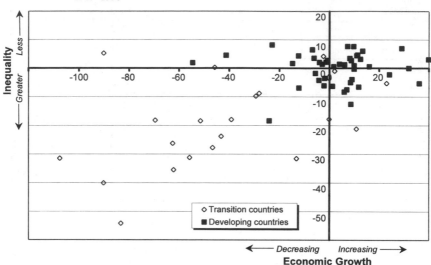

Source: Martin Ravallion and Shaohua Chen, "What Can New Survey Data Tell Us About Recent Changes in Distribution and Poverty?" *World Bank Economic Review* (Washington, DC: World Bank 1997).

Notes: Inequality = Gini index x –100; Economic growth = log of change in average household consumption x 100.

▨ Sustainable Development

Economic growth is necessary to achieve more fundamental human goals, at least in poor countries, but alone is not sufficient. The concept of *sustainable development* has emerged to incorporate other aspects of development in addition to economic growth. Advocates of international justice, environmental protection, peace, sustainable population growth, democracy, and human rights have increasingly come to see that their goals are interlinked. Several examples follow:

- There is no way to save the rain forests of Brazil without dealing with the land hunger of poor Brazilians.
- There is no way to reduce rapid population growth in developing countries without improving living standards, especially for girls (see Chapters 9 and 10).
- There are no durable solutions to poverty and hunger in the United States without social peace, broader democratic participation, and a shift to economic patterns that will be environmentally sustainable.

Bread for the World defines sustainable development in terms of four interconnected objectives: providing economic opportunity for poor people, meeting basic human needs, ensuring environmental protection, and enabling democratic participation (BFW Institute 1995). These concepts, and those from other chapters in this book, are reflected in the following policy suggestions.

■ ANTIPOVERTY POLICIES IN A GLOBAL ECONOMY

Public policies aimed at reducing poverty fall into two broad categories: (1) creating appropriate, effective guidelines for markets; and (2) collecting and allocating public resources, especially for investment in human resources.

The reality of achieving effective antipoverty policies is, of course, much more difficult than the assertion. Just as some actors in the marketplace can take advantage of their economic power, they and other powerful political actors can sway policies to their own self-interest, whether at the local, national, or international level. Meanwhile, poor people, whose well-being is the strongest evidence of whether policies are effective, often lack political access or clout. In addition, they and their allies are often unclear or divided on issues of national and international economic policy.

But as the global, knowledge-based, market economy reaches into the far corners of our planet, people of goodwill have only one option: to help draft and implement policies that will direct a sizable portion of this economy toward creating income-earning opportunities for poor people. The most important areas for policies to help reduce poverty include investing in people; sustaining agriculture and food production; creating a framework for sustainable development; and targeting international financing (BFW Institute 1997; World Bank 1997a).

▒ Investing in People

Health care and nutrition. Investments in basic health care and improved nutrition yield huge dividends. Healthy children learn better. Healthy adults work better. Improved health care begins with greater attention to basic public health measures: nutrition education, clean water and adequate sanitation, vaccination against infectious diseases, prevention of AIDS, distribution of iodine and Vitamin A capsules, and simple techniques of home health care. Delivery of these services can be relatively inexpensive, especially in developing countries, where village women with minimal training can be employed. These basic services should have priority over urban hospitals and specialized medical training.

In some instances, public health training can be delivered in conjunction with supplemental feeding programs such as the Special Supplemental

Food Program for Women, Infants, and Children (WIC) in the United States, or the Integrated Child Development Services in India.

Education. Investments in basic education complement those in health care and improved nutrition and yield huge payoffs in both developing and industrialized nations. Better education for youth, especially girls, leads to improved health awareness and practices for their families on a life-long basis. Cognitive and other skills improve productivity, enable better management of resources, and permit access to new technologies. They also enhance participation in democracy.

A study of ninety-eight countries for the period from 1960 to 1985 showed GDP gains up to 20 percent resulting from increases in elementary school enrollment, and up to 40 percent resulting from increases in secondary enrollment. In allocating educational resources, the highest payoff is for elementary education, because it reaches the most children (Fiske 1993).

In the United States, dramatic improvement has followed investments in Head Start, which provides preschool education and meals for low-income children; and Job Corps, which provides remedial and vocational training for disadvantaged youth (see Figure 8.6).

Agriculture and Food Production

Access to land. Widespread land ownership by small farmers usually contributes directly to food security and improved environmental practices. The more successful land reform programs, as in Korea and Taiwan, have provided at least minimum compensation to existing landlords.

Figure 8.6   Virtuous Circle

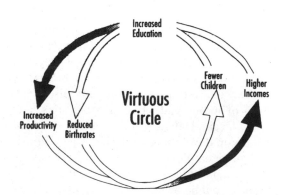

Source: Bread for the World (BFW) Institute, *Hunger 1995: Causes of Hunger* (Washington, DC: BFW Institute, 1994), p. 56. Used by permission of Bread for the World.

Equitable prices for farm produce. Thriving agriculture is basic to successful development in most of the poorest nations. Much of new savings must be accumulated within agriculture, since it is such a large share of the economy. Such savings are important for increasing agricultural productivity and for helping finance rural, nonfarm businesses. Also, as their incomes rise, farmers expand their purchases of consumer goods, providing an important source of nonfarm employment.

In many developing countries, state-run marketing boards have taxed agriculture by setting farm prices very low and retaining for the government a large share of the value from farm exports. Meanwhile, the United States and the European Union have supported their own farmers in ways that generate surplus crops. They also subsidize exports of these crops, driving down prices around the world. Developing country farmers, who usually are not subsidized, cannot match the low prices. Agriculture falters and with it the whole process of development. Both rich country export subsidies and developing country discrimination against agriculture should be phased out as quickly as possible.

◼ Creating a Framework for Sustainable Development

Access to credit. Equitable access to credit for small farmers and small businesses is probably the highest priority for the allocation of domestic savings or outside investment. Training in resource and business management is often part of successful credit programs.

Most informal economic activity results from the efforts of small entrepreneurs who cannot find a place in the formal economy. If they have access to good roads, markets, and credit, small farmers and small business people can create their own new income-earning opportunities in market economies.

Adequate physical infrastructure. Creating and maintaining an adequate physical infrastructure is essential to a viable, expanding economy. Important for rural areas are farm-to-market roads and food storage, both oriented to domestic production and, if appropriate, exports. For all areas, safe water, sanitation, electricity, and communications networks are needed.

Stable legal and institutional framework. Sustainable development requires a stable legal framework. This includes assured property titles, enforceable contracts, equitable access to courts and administrative bodies, and accessible information networks.

Stable currency and fiscal policies. Neither domestic nor international investors, including small farmers and microentrepreneurs, are likely to invest in countries in which the political or economic environment is unsettled. High inflation or continuing trade deficits, which often go together, discourage needed investments and may even drive out domestic savings.

Twenty-nine African countries undertook structural adjustment programs during the 1980s. These reforms have often fallen heavily on poor people, as social service programs were frozen or cut back to help balance national government budgets. Other measures to reduce budget or trade deficits—more progressive taxes, cuts in military spending, and curtailing luxury imports—have not always been pursued with equal vigor. Economic reform programs, which will continue to be necessary, should be revamped to reduce the costs to, and increase benefits for, poor people.

Effective, progressive tax systems. Effective progressive tax structures are key to sustainable financing for investments in human resources and infrastructure. Taxes based on ability to pay are also key to stabilizing or reducing wide disparities in income distribution in both rich and poor countries. Such tax systems are difficult to enact where wealth and political power are controlled by a small minority.

Incentives for job-creating investments. The Asian countries that have grown so rapidly have all placed emphasis on labor-intensive exports—some in joint ventures with overseas partners, some with investments solicited from abroad, but many with subsidies from within their own economies. This is a distinct departure from their more general commitment to following market signals.

A primary target for job-intensive investments will be processing operations for primary products—whether for domestic consumption or for crops and minerals now being exported.

▩ Targeting International Financing

The importance of international financing is well covered in Chapter 7. The only addition here is to emphasize the distinction between overall growth and growth that will create more and better opportunities for poor people.

■ SUMMARY AND CONCLUSION

Poverty—the lack of resources or income to command basic necessities—is the condition of about one-fifth of the world's population, or about one-third of the people in developing countries. The absolute number of poor people has increased slowly in recent years, while their proportion has declined slightly, with considerable regional variation.

Economic growth is essential to overcoming poverty, but it alone is not sufficient. Relatively egalitarian distribution of national income among and within households also matters greatly. Creating 2 billion good jobs or business opportunities is the biggest single challenge for this generation. The economic and policy tools to generate relatively equitable growth have been successfully demonstrated in recent years, particularly in East and Southeast Asia. Meanwhile, some of the worst effects of poverty have been, and should continue to be, offset by public and private interventions—infant mortality and overall hunger have declined, and literacy and longevity have increased in many instances, even in the face of continued poverty.

Adapting these tools and programs to particular circumstances, especially in Africa, is of utmost concern to everyone. In an increasingly global economy, the well-being and security of each person or community or nation is inescapably linked to that of every other.

The ingredient in shortest supply to overcome poverty appears to be the political will to do so.

■ QUESTIONS

1. Are you more inclined to measure poverty in terms of absolute income, income distribution, or the capacity to reach more fundamental goals? If the latter, what would be your list of goals?

2. Which of the antipoverty policies do you think are the most beneficial?

3. What would you consider a reasonable goal for the ratio between the top and bottom income groups within an economy? Within a business firm? What policies would be necessary to move toward this goal?

4. Should government policies encourage the redistribution of income? If so, to what extent?

5. Are you as optimistic as the author that poverty can be overcome?

6. Do you concur that the well-being of everyone is "inescapably linked"?

7. Do you agree with the author that the principal missing ingredient in overcoming hunger is "political will"?

■ SUGGESTED READINGS

Birdsall, Nancy, Thomas Pinckney, and Richard Sabot (1996) "Why Low Inequality Spurs Growth: Savings and Investment by the Poor," Inter-American Development Bank Working Paper Series, No. 327. Washington, DC: IDB.
Birdsall, Nancy, David Ross, and Richard Sabot (1995) "Inequality and Growth Reconsidered: Lessons from East Asia," *World Bank Economic Review* 9, no. 3.

Bread for the World Institute (annual). Washington, DC: BFW Institute.

Drêze, Jean, and Amartya Sen (1989) *Hunger and Public Action*. Oxford: Clarendon Press.

Fiske, Edward B. (1993) *Basic Education: Building Block for Global Development*. Washington, DC: Academy for Educational Development.

United Nations Development Programme (annual) *Human Development Report*. New York: Oxford University Press.

World Bank (annual) *World Development Report*. New York: Oxford University Press.

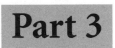

Part 3

Development

Population and
Migration

Ellen Percy Kraly

Coming to grips with the implications of current population trends is an extremely important dimension of global studies. The process is neither easy nor comforting, because significant population increase is an inevitable characteristic of the global landscape in the first fifty years of the twenty-first century. It is critical that students interested in global studies and issues should appreciate both the causes of population growth and consequences of population change for society and the environment. Such an appreciation will serve in developing appropriate and effective responses to population-related problems emerging globally, regionally, and locally.

This chapter seeks to contribute to the understanding of the interconnections among population change, environmental issues, and social, economic, and political change in both developing and developed regions of the world. Because population growth has momentum that cannot be quickly changed, it is important to begin by considering fundamental principles of population or demographic analysis and to place recent global and regional population trends in historical perspective. The chapter then examines the widely divergent philosophical and scientific perspectives on the relationships among population, society, and environment that have pervaded visions of the future. Debates on the implications of current growth have also influenced discussions about routes for population policy. In the next section, global effects of population redistribution, urbanization, and

Assistance with illustrations was provided by Jennifer Critchley and Patrick Rowe, Department of Geography, Colgate University.

international migration are discussed. The chapter concludes by considering global dimensions of population policies targeting growth and international population movements.

■ PRINCIPLES AND TRENDS

▣ Population Concepts and Analysis

Demography is the study of population change and characteristics. A population can change in size and composition as a result of the interplay of three demographic processes: fertility, mortality, and migration. These components of change constitute the following population equation:

$\Delta P = (+)$ births $(-)$ deaths $(+)$ in-migration $(-)$ out-migration;
where ΔP is population change between two points in time

On the global level, the world's population grows as the result of the relative balance between births and deaths, often called natural increase. The U.S. population is currently increasing at about 0.9 percent per year; natural increase accounts for about two-thirds, and net international migration constitutes about one-third of this relatively low level of population growth.

Many people seeking routes to sustainable development advocate a cessation of population growth often referred to as zero population growth (ZPG). When viewed from a short-run perspective, ZPG means simply balancing the components of the population equation to yield no (zero) change in population size during a period of time. Population scientists, however, usually consider ZPG in a long-term perspective by considering a particular form of a zero-growth population: a stationary population is one in which constant patterns of childbearing interact with constant mortality and migration to yield a population changing by zero percent per year. In such a case, fertility is considered replacement fertility because one generation of parents is just replacing itself in the next generation. In low-mortality counties, replacement-level fertility can be measured by the total fertility rate and is approximately 2.1 births per woman to achieve a stationary population over the long run.

It takes a relatively long time, perhaps three generations after replacement fertility has been achieved, for a population to cease growing on a yearly basis. Large groups of persons of childbearing age, reflecting earlier eras of high fertility, result in large numbers of births even with replacement-level fertility. Hence, an excess of births over deaths occurs until these "age structure" effects work themselves out of the population. This is known as the momentum of population growth. To illustrate, it is estimated that if the world's populations achieved replacement-level fertility immediately, an extremely unrealistic scenario, global population would continue to grow from 5.3 billion in 1990 to 7.4 billion in 2050, an increase of 40 percent (Lutz 1994).

■ Historical and Contemporary Trends in Population Growth

The world's population was estimated to be 5.8 billion in 1996 and increasing at approximately 1.5 percent per year (USBC 1996). These data represent a cross-sectional perspective on population characteristics—a snapshot that fails to capture the varying pace of population change worldwide and regionally. Over most of human history, populations have increased insignificantly or at very low annual rates of growth, with local populations being checked by disease, war, and unstable food supplies. Between the sixteenth and the eighteenth centuries, population growth appeared to become more sustained as a result of changes in the social and economic environment: improved sanitation, more consistent food distribution, improved personal hygiene and clothing, political stability, and the like.

The world's population probably did not reach its first billion until just past 1800. But accelerating population growth during the nineteenth century dramatically reduced the length of time by which the next billion was added. According to the United Nations (UNDESIPA 1994), world population reached:

- 1 billion in 1804
- 2 billion in 1927 (123 years later)
- 3 billion in 1960 (33 years later)
- 4 billion in 1974 (14 years later)
- 5 billion in 1987 (13 years later)

If the current estimated annual rate of global population growth of 1.5 percent is maintained, the world's population will double to nearly 12 billion in just over forty-six years. A closer look at the pattern of population change suggests, however, that this is an unlikely scenario. Population data for years since 1950 are shown in Figure 9.1. Between 1950 and 1996, the world's total population increased from 2.6 billion to 5.8 billion.

The difference in height of the bars in Figure 9.1 also reveals the momentum of population growth that results in continued additions to the world's population, albeit in decreasing numbers: between 1985 and 1990, approximately 85 million persons were added to the world's population each year; since 1990, the annual increase is estimated at 82 million.

It is important to note, however, that in spite of these large additions to the world's population, the rate of population growth is *decreasing*. The

Figure 9.1   World Population for Development Categories, 1950–1996

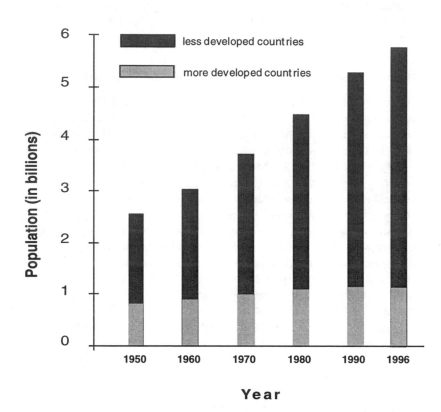

Source: U.S. Bureau of the Census, International Data Base.

average annual rate of global population growth reached an all-time high of about 2.2 percent between 1962 and 1964. Since that time, the pace of the growth of the world's population has decreased to the current rate of approximately 1.5 percent per year.

Patterns of population growth differ significantly between more and less developed regions of the world. Table 9.1 provides greater geographic detail and summarizes population size and distribution for major regions of the world for selected years since 1950.

Dramatic shifts in the geography of world population have occurred during the past five decades and, as discussed below, are expected to continue well into the future. In 1950, just over two-thirds of the world's population was located in less developed countries; by 1996, this proportion had increased to four-fifths. Asian countries comprise over half the world's

Table 9.1   World Population by Geographic Region and for More and Less Developed
Countries, 1950–1996

	1950	1960	1970	1980	1990	1996
Region	Population (in millions)					
World	2,556	3,039	3,706	4,458	5,282	5,772
Less developed countries	1,749	2,129	2,703	3,377	4,139	4,601
More developed countries	807	910	1,003	1,081	1,142	1,171
Africa	29	283	360	470	624	732
Sub-Saharan Africa	185	227	289	379	504	594
Asia	1,411	1,685	2,113	2,601	3,123	3,428
Latin America and Caribbean	166	218	285	362	443	489
Europe	572	639	703	750	789	800
North America	166	199	226	252	277	295
Oceania	12	16	19	23	27	29
Region	Percent Distribution					
World	100.0	100.0	100.0	100.0	100.0	100.0
Less developed countries	68.4	70.1	72.9	75.8	78.4	79.7
More developed countries	31.6	29.9	27.1	24.2	21.6	20.3
Africa	1.1	9.3	9.7	10.5	11.8	12.7
Sub-Saharan Africa	7.2	7.5	7.8	8.5	9.5	10.3
Asia	55.2	55.4	57.0	58.3	59.1	59.4
Latin America and Caribbean	6.5	7.2	7.7	8.1	8.4	8.5
Europe	22.4	21.0	19.0	16.8	14.9	13.9
North America	6.5	6.5	6.1	5.7	5.2	5.1
Oceania	0.5	0.5	0.5	0.5	0.5	0.5

Source: U.S. Bureau of the Census, 1994, Table A-1.

population; over one-fifth, 21 percent, of the global village lives in China
and another 16 percent in India (data not shown in Table 9.1). Africa's
share has increased from just under one-tenth in 1950 to about 13 percent
of the current population. European populations constituted 14 percent of
the world's population in 1996, a decline from 22 percent in 1950. Western Hemisphere regions—North America, Latin America, and the
Caribbean—include approximately 14 percent of the world's population.

Population growth is fueled by levels of fertility, mortality, and net
migration. The rapid population growth that occurred in the post–World
War II era reflected significant declines in mortality resulting in large part
from the transfer of medical technology and public health advances from
more to less developed countries.

The total fertility rate measures the average number of births per
woman of childbearing age and is a strong indicator of overall population
growth. In 1996, the total fertility rate for the world as a whole was 2.9

births per woman, representing a significant decline from 4.2 in 1985. Fertility in more developed countries has been below replacement for some time and is currently estimated at 1.6 births per woman. In developing countries, the rate has dropped from 4.7 in 1985 to 3.3 in 1996. Much of this decline is weighted by the aggressive fertility control campaign in China and significant declines in fertility throughout Southeast Asia and in Latin America. Total fertility in India has also declined but less dramatically, from 4.3 in 1985 to 3.2 currently. Total fertility in Africa, particularly sub-Saharan Africa, while declining somewhat in the past decade, remains strikingly high: 5.5 births per woman in 1996. Figure 9.2 provides a cartographic view of current levels of fertility for countries of the world.

■ PERSPECTIVES ON THE CAUSES AND CONSEQUENCES OF POPULATION CHANGE

Reflections on the relationship between population and society can be found in the early history of many cultures. In early Greece, Plato wrote about the need for balance between the size of the city and its resource base; Confucianism emphasized the social and economic advantages of large families. Concern about the implications of population growth for social progress became a focus of social theory in the nineteenth century and continues in contemporary debates on global effects of current levels of population growth.

■ Debates on Population Growth

Certainly the most influential statement concerning the sources and implications of population growth was that of Thomas Malthus in his *Essay on the Principle of Population;* the first edition, published in 1798, was followed by several revisions (Malthus 1878). In his essay, Malthus was reacting to mercantilist philosophy that pervaded eighteenth-century European thought and emphasized the value of large and increasing populations for economic growth and prosperity. Malthus offered a negative perspective on the consequences of population growth, arguing that the cumulative, or "geometric," nature of population growth will outpace the increase in food supply (which Malthus believed increased arithmetically) and would result in starvation, poverty, and human misery. Because of the instinct of humans to reproduce, population growth can only be halted through the so-called positive checks—rising mortality as a result of famine, war, and epidemics. Malthus, a clergyman, opposed contraception and advocated delayed marriage and abstinence, which he called "preventive" checks. Neo-Malthusians, in contrast, recognize the importance of

Figure 9.2 Total Fertility Rates, 1996

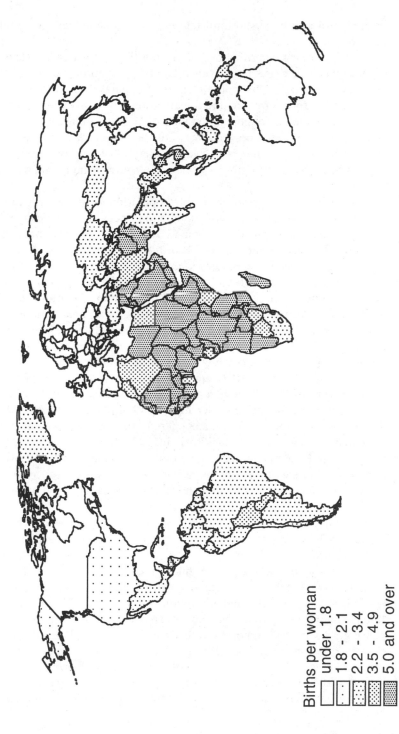

Births per woman
under 1.8
1.8 - 2.1
2.2 - 3.4
3.5 - 4.9
5.0 and over

Source: U.S. Bureau of the Census, *World Population Profile* (Washington, DC: Government Printing Office, 1996), Table A-8.

birth control as a means to limit family size and hence reduce population growth.

Among the early critics of Malthusian thought was Karl Marx, who argued that there exists no universal law of population such that Malthus had forwarded; rather that consequences of population growth derive from the particular form of economic organization within a society. According to Marx, capitalism, not population growth, resulted in increasing levels of poverty. Thus, population growth is not the source of social, economic, and environmental problems; instead, society must be reorganized to provide an equitable distribution of access to resources and the means of production.

These historical perspectives on population hold a true vitality for contemporary debates concerning population issues and policies. Two lines of thought have dominated the population debate in the past three decades and might be crudely labeled *neo-Malthusian* and *cornucopian*. The neo-Malthusian perspective continues to emphasize the problem of population growth as the primary obstacle to sustaining the ecological balance of planet earth by leading to natural resource depletion, pollution, and loss of biodiversity. Cornucopian perspectives, on the other hand, emphasize the role of technological innovation and market forces, which through pricing effects will manage the use of natural resources. From this vantage point, population growth holds potential for solving global problems through increased economic productivity and capacity for technological progress.

The work of Paul Ehrlich, a biologist and ecologist, has been very influential in the contemporary debate on global population growth and is representative of the neo-Malthusian perspective. Ehrlich, in collaboration with other scholars (Ehrlich and Ehrlich 1992; Holdren and Ehrlich 1974), has developed a general model that views environmental impact as a function of the interactions among population size and structure, consumption behaviors, and technology. A large body of empirical research has emerged to assess the nature of relationships among these dimensions of environmental impact. From this perspective, Ehrlich and others conclude that world population growth has significant and globally far-reaching impacts on ecological sustainability.

The cornucopian perspective on population growth is held by some economists and others who recognize the potential of technology and market forces to solve social and environmental problems. The scholarship of Julian Simon (1990) represents an extreme statement of this view by espousing a positive vision of the future based on a record of human history characterized by technological progress, human ingenuity, and an overarching trend toward rising standards of living and quality of life. A contribution to this perspective was Ester Boserup's (1981) research on long-term agricultural change, which shows that historical increases in population density were followed by technological change in farming techniques and growth in food production.

Because of the dramatically different visions of society and nature inherent in these two perspectives, the debate between the negative and positive consequences of population growth has always been energetic, animated, and often contentious. It is therefore helpful to many concerned students of population studies that a third perspective on population, society, and environment linkages has emerged in the past few decades. This perspective has been labeled by some as *structuralist* (see Harper 1995).

Structuralist perspectives, which borrow from Marxist theory, consider population characteristics, including population growth, poverty, food supplies, and environmental problems, as outcomes of broader social structural processes and institutions. Thus, population growth, specifically high fertility, is more a consequence than a cause of slow economic development and restricted social and economic opportunities. Barry Commoner, also a biologist, emphasizes poverty and the lack of options for women as factors in continued high fertility in developing societies; he stresses the greater relative significance of high levels of energy consumption and waste in rich northern countries for the future global environment (Commoner 1992).

From these disparate perspectives, Charles Harper (1995) sees an "emerging consensus" concerning the consequences of population growth. This perspective conceptualizes population growth as both cause and consequence of social, economic, and environmental processes confronting the world, nations, and local communities. Hence, the reduction of poverty, the improvement of life chances and of the status of women, the increasing sustainability in food production, the improvement in water quality, etc., all become important strategies for reducing population growth. Also recognized is the advantage of slowing and ultimately ceasing world and regional population growth in order to more effectively improve standards of living, stabilize food supplies, and halt environmental degradation (see National Research Council 1986).

Helping to chart the progress toward low population growth in societies is the model of the demographic transition that was developed initially as a description of population growth patterns in Europe and North America in the nineteenth and early twentieth centuries. The model became linked to theories of modernization to predict how population growth would proceed throughout the developing world. As societies underwent industrialization and urbanization, with all the concomitant social changes, death rates would fall as they had in the Western societies, followed with a lag by declines in fertility. During the lag, population growth would occur until norms and values concerning the need for large families were replaced by small family ideals.

The demographic transition model, shown schematically in Figure 9.3, has been widely criticized, retested, and revised. A major limitation of the theory is the view that the experiences of non-Western societies will mirror or converge with those of Europe and North America. Revisions of the demographic transition model have a clearer understanding of how family size is influenced by cultural beliefs and by gender, particularly educational opportunities for girls. Moreover, studies in sub-Saharan Africa and South Asia have shown the importance of understanding the contributions of children to the well-being of the family and its kin. Young children in developing societies may be a source of wealth as household labor and family prestige, and older children may provide old-age support and continuation of the family lineage (Caldwell 1982).

■ EXPECTATIONS ABOUT FUTURE POPULATION GROWTH

Theories of population change guide analyses of future population growth, usually in the form of population projections. Most demographers are quick to state that population projections are not predictions but represent a calculation of future population size based on a set of assumptions or variants. Shown in Figure 9.4 are population estimates and projections prepared by the United Nations Population Division for the years 1950–2050. The world totals reflect the sum of projections conducted for 228 countries or areas; the "fan" of population figures represents the three projection variants for the projection period 1990–2050.

Figure 9.3   The Demographic Transition Model

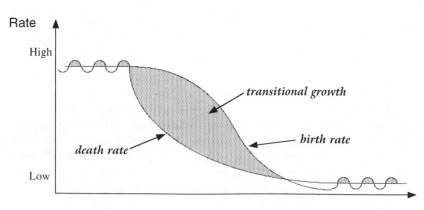

Figure 9.4 Projected World Population, 1950–2050

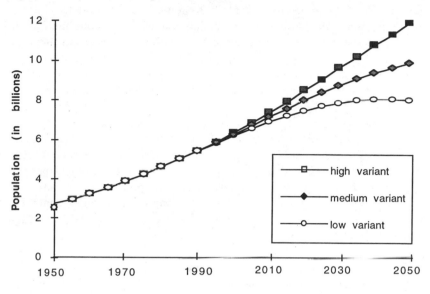

Year

Source: UNDESIPA, "The Sex and Age Distribution of the World's Populations" (New York: United Nations, 1994).

Three projection variants are displayed, high, medium, and low, which in turn reflect trends in fertility. In the high variant, fertility is assumed to reach a total fertility rate of 2.6. In the low variant, fertility is assumed to reach 1.6 births per woman. The medium variant targets fertility at approximately replacement level, 2.1 births per woman. In each variant, the pace of fertility change is related to the social and economic characteristics of the individual country. In all three projections, mortality rises, although at a slow rate of improvement in lower-mortality countries. Hence, the series as a whole assumes convergence among countries to common fertility levels and mortality levels, generally embodying a demographic transition model.

The high-fertility scenario results in growth in world population from 5.3 to nearly 11.9 billion between 1990 and 2050, well over a doubling of the world's population. The low scenario projects an increase to 7.9 billion in 2050, an increase of nearly 50 percent. Thus, even with convergence of fertility levels to well below replacement (total fertility rates of 1.6), the world's population will continue to grow as a result of the momentum of population growth. The medium variant, which models trends toward replacement fertility, results in an increase to 9.8 billion in 2050, an increase of 85 percent over the 1990 population.

Shifts in the geographic distribution of the world's population are evident in all of the projections. Table 9.2 shows these geographic changes for the medium population scenario. Asian countries will continue to hold the largest share of the world's population, although decreasing from 60 percent to approximately 58 percent. It is important to consider two Asian countries in particular, China and India, whose population policies are considered later. The population of China is projected to increase from 1.2 billion in 1990 to 1.6 billion in 2050; India's population is projected to grow from 851 million to 1.6 billion in 2050, thus equaling China. A full one-third of the world's population may be living in one of these two countries by halfway through the next century (UNDESIPA 1994). African countries, primarily in sub-Saharan regions, will increase from 12 percent to 22 percent of the world's population, and the population of Europe is expected to decline from 14 percent to 7 percent of the world's population. Thus, we can expect significant shifts in world geography of population based on these projections.

Table 9.2   Projections of the World Population by Geographic Region and for More and Less Developed Countries, 1996-2050

	1990	2000	2025	2050
Region	Population (in millions)			
World	5,285	6,158	8,294	9,833
More developed countries	1,143	1,186	1,238	1,208
Less developed countries	4,141	4,973	7,056	8,626
Africa	633	832	1,496	2,141
Sub-Saharan Africa	514	686	1,286	1,885
Asia	3,186	3,736	4,960	5,741
Europe	722	730	718	678
Latin America and Caribbean	440	524	710	839
North America	278	306	370	389
Oceania	0	0	0	0
Region	Percent Distribution			
World	100.0	100.0	100.0	100.0
More developed countries	21.6	19.3	14.9	12.3
Less developed countries	78.4	80.7	85.1	87.7
Africa	12.0	13.5	18.0	21.8
Sub-Saharan Africa	9.7	11.1	15.5	19.2
Asia	60.3	60.7	59.8	58.4
Europe	13.7	11.9	8.7	6.9
Latin America and Caribbean	8.3	8.5	8.6	8.5
North America	5.3	5.0	4.5	4.0
Oceania	0.0	0.0	0.0	0.0

Source: UNDESIPA, "The Sex and Age Distribution of the World's Populations" (New York: United Nations, 1994).

■ SOCIAL AND ENVIRONMENTAL DIMENSIONS OF POPULATION REDISTRIBUTION AND MIGRATION

The shift in the world's population from developed to developing regions reflects differences in fertility and mortality as well as migration to and from these regions. Moreover, the movement of persons within and between countries is both a cause and consequence of social, economic, political, and environmental factors.

Geographic mobility is the general concept covering all types of human population movements. Migration is generally considered to refer to moves that are permanent or longer term; internal migration within a country is distinguished from international population movements; international migration to a country is *immigration,* and international migration from a country is *emigration.* Reasons for moving are often included in migration concepts, such as labor migration, refugee migration, and seasonal migration. Internal and international migration are processes that are increasingly linked through the geographic and social dimensions of global economic development.

■ Internal Migration and Urbanization

A corollary of the demographic transition model is growth in the size of cities as well as increasing proportions of populations living in cities and metropolitan areas, that is, urbanization. The UN estimates that 45 percent of the world's population was living in urban areas in 1994; differences between more and less developed countries are dramatic, 75 percent and 37 percent, respectively. It is important to remember that while some of the largest cities in the world are found in developing countries—for example, São Paulo, Brazil, with a metropolitan population over 16 million; Mexico City with nearly 16 million; and Shanghai and Bombay each with nearly 15 million—the majority of populations in these societies currently live in rural areas (UNDESIPA 1995).

Given population growth and rural-to-urban migration, however, urbanization is expected to increase throughout regions of the world. By the year 2050, for example, the level of urbanization in more developed countries is expected to increase to 84 percent and in less developed countries even more steeply to 57 percent, a dramatic shift in patterns of residence and economic activity (UNDESIPA 1995).

The causes of urban growth have varied among regions and during different historical periods. In Western societies, urbanization was fueled in large part by technological change in both agricultural and industrial sectors, resulting in both a push from rural communities and a pull to emerging industrial centers (Harper 1995). In developing societies, rural-to-urban migration is driven by many factors, including increasing population density (caused by high fertility rates) in rural areas, environmental degradation from practices such as overgrazing, and the hope for gainful employment in urban areas. The pull of cities in many developed countries exists in the form of the hope for employment and higher wages rather than prearranged jobs. As a result, levels of unemployment in cities in developing countries are very high. Evidence of underemployment is shown in the large numbers of persons, including many children, attempting to earn livelihoods in what has been called by some the informal economy—for example, street vendors, curbside entertainers, and newspaper boys and girls.

■ International Population Movements

One of the most visible manifestations of globalization is the increasing scale of international population movements throughout all regions of the world. According to Castles and Miller,

> large-scale movements of people arise from the accelerating process of global integration. Migrations are not an isolated phenomenon: movements of commodities and capital almost always give rise to movements of people. Global cultural interchange, facilitated by improved transport and the proliferation of print and electronic media, also leads to migration. (Castles and Miller 1993: 3)

The consequences of international population movements for both sending and receiving nations and communities will have significant implications for emerging global issues.

International population movements involve an increasingly large number of countries, both as sending and receiving regions, and include many different types of migrants. Countries in both Western and Eastern Europe have been faced with large numbers of persons seeking political asylum from both European regions as well as geographically distant sources, including East Africa and Southeast Asia; significant labor migration flows have emerged between South and Southeast Asia and oil-producing regions of the Middle East and throughout the Asia-Pacific region; the United States has grappled with issues concerning large numbers of undocumented migrants drawing from Mexico and many other source countries; refugee migration is recurringly characteristic of political and environmental change throughout Africa. Emerging trends include the international circulation of professional workers, sometimes called professional transients; and in many international migration flows, including labor migration, the proportion of women is increasing.

The significance of the scale of refugee migration and displaced populations in global population issues cannot be understated. The international definition of a refugee is a person who,

> owing to a well-founded fear of being persecuted for reasons of race, relation, nationality, membership of particular social group(s) or political opinion is outside the country of his nationality and is unable to or owing to such fear is unwilling to avail himself of the protection of that country; or who, not having a nationality and being outside the country of his former habitual residence . . . is unable or unwilling to return to it. (UNHCR 1995: 256)

In 1995, the Office of the UN High Commissioner for Refugees (UNHCR) identified more than 27 million persons throughout the world who were of concern to the organization, an increase of 10 million in just four years. Of this extraordinary number, 14.5 million are recognized as refugees living in asylum in other countries; the remainder are persons who are internally displaced within their own countries for complex political, economic, and environmental reasons, and persons outside their home country in refugee-like situations. Table 9.3 displays refugees by broad category and geographic region estimated by the UNHCR for 1995. Many refugee settlements or camps have existed for many years; some refugees have been repatriated to their homelands—for example, Guatemalans who had sought refuge in Mexico and Muslims who had fled Myanmar (formerly Burma); others, including many Vietnamese during the 1970s and 1980s, have been permanently resettled in other countries such as Canada, Australia, and the United States. The majority of the world's refugees are women and children, whose voices are often not heard in discussions about programs to aid and resettle refugees (UNHCR 1995).

■ POPULATION POLICIES

A direct or explicit population policy is a "strategy for achieving a particular pattern of population change" (Weeks 1996: 496). According to the UN survey of government policies, approximately one-third of all countries have policies or programs to reduce population growth; a significant number of smaller countries, including France, Cyprus, Greece, Hungary, and Gabon, have goals to increase growth (UNDESIPA 1995). Many more countries have indirect population policies that, while not targeting population growth, have clear implications for either mortality, fertility, or migration. The United States, for example, has not yet adopted a formal statement of goals concerning national population growth, but does have

Table 9.3   Refugees and Other Types of Persons of Concern to the UNHCR, by Geographic Region, 1995

Region	Total	Refugees	Returnees	Others of Concern	Internally Displaced
	Population (in thousands)				
World	27,819	14,889	3,983	3,524	5,423
Africa	11,816	6,752	3,084	7	1,973
Asia	7,922	5,018	832	310	1,762
Latin America and Caribbean	185	109	67	1	8
Europe	6,520	1,867	0	2,963	1,680
North America	926	681	0	244	0
Oceania	51	51	0	0	0
Region	Percent Distribution				
World	100.0	100.0	100.0	100.0	100.0
Africa	43.1	45.3	77.4	0.2	36.4
Asia	28.9	33.7	20.9	8.8	32.5
Latin America and Caribbean	0.7	0.7	1.7	0.0	0.0
Europe	23.8	12.5	0.0	84.1	31.0
North America	3.4	4.6	0.0	6.9	0.1
Oceania	0.2	0.3	0.0	0.0	0.0

Source: UNHCR, *The State of the World's Refugees 1995* (Oxford: Oxford University Press, 1995).
Note: Totals may not add up due to rounding.

a long-standing policy for the permanent resettlement of immigrants and refugees, which in turn results in net additions to the population through international migration.

▪ International Efforts to Reduce Population Growth

International population conferences bring government delegations and representatives of nongovernmental organizations (NGOs) together to discuss goals concerning population and to develop strategies for achieving those goals. The most recent conference was held in Cairo in 1994. In each of these gatherings, there has been a general recognition, though not universal agreement, that (1) rapid population growth fueled by high fertility poses a challenge to economic development in less developed countries; (2) mortality should be reduced regardless of the effect on population growth; and (3) international migration is an appropriate arena of national policy and control (Weeks 1996). Important shifts in thinking about population growth, however, have occurred during the past three decades that have had implications for the agenda of population policies and programs within countries. These are discussed in the next section.

The 1974 World Population Conference in Bucharest produced the first formal expression of a world population policy. The "World Population Plan of Action," however, embodied a wide range of perspectives on the ways to reduce population growth within developing societies. Some countries, notably the United States, advocated fertility control, specifically family planning programs, to reach population growth targets. Other countries, primarily in the developing world, emphasized the role of development in leading to fertility decline (hence, "development as the best contraceptive"). The 1984 International Population Conference in Mexico City found the United States reducing its support for family planning (which was linked in turn to the Reagan administration's views on abortion) and identifying population growth as having little hindrance on economic and social development. Many developing countries by this time, however, had instituted family planning programs in an effort to slow the retarding effects of rapid population growth on improving standards of living and educational levels and reducing mortality (Weeks 1996).

The 1994 International Conference on Population and Development recognized the global dimensions of population change. The "Programme of Action" identifies the connections among population processes, economic and social development, human rights and opportunities, and the environment, thus shifting attention away from targets concerning population growth to goals concerning sustained development, reduction of poverty, and environmental balance. The central role of women in the goals and programs to achieve sustainable development is underscored in the final report: "The key to this new approach is empowering women and providing them with more choices through expanded access to education and health services, skill development and employment, and through their full involvement in policy- and decision-making processes at all levels" (UNDESIPA 1995: 1).

The emphasis on connecting population issues to the status of women represents a significant forward step in embedding population analysis in broader discussions of the quality of human life and the balance between society and environment at local, national, and global levels.

◼ National Population Policies

Most countries that have formal population policies seek to reduce population growth by reducing fertility. Beginning in the 1950s, providing contraception through family planning programs was initiated in many developing countries—often with significant contributions from developed countries, NGOs and foundations, and, in subsequent decades, international organizations such as the UN Fund for Population Activities (UNFPA). Increasingly, policies aimed at fertility reduction have encompassed broader perspectives on population dynamics, incorporating goals to increase the status of women through better health, enhanced educational and employment opportunities, greater access to credit, etc.

The record of family planning programs has been variable throughout the developing world. India has maintained a national family planning program since 1952 that has met with fertility trends that vary significantly throughout regions within the country and between rural and urban areas. In 1973, Mexico instituted a national policy to reduce population growth, with focus on the reduction of fertility through maternal and child health programs, family planning services, sex education, and population information programs. China's one-child policy began in 1971 as a set of policy goals concerning later marriage, longer intervals between births, and fewer children, with the one-child policy implemented in 1979. The Chinese program combined the provision of contraception with social and economic incentives and resulted in significant overall fertility decline; there was a more than 40 percent decline in the birthrate during a ten-year period. The Chinese policy has been criticized harshly, however, for being coercive and for leading to selective abortion and infanticide of girls.

■ International Migration and Refugee Policies

The UN Declaration of Human Rights recognizes the basic right of people to leave their homeland. The converse of this right to emigrate, however, is not recognized—that is, nation-states have the sovereign right to control the entry of nonnationals into their territory. Nearly all countries have clear policies concerning international migration and travel. In fact, very few countries continue to allow international migration for permanent resettlement, including the United States, Canada, Australia, and New Zealand, often considered the traditional immigrant receiving countries. A much larger range of countries provide humanitarian assistance to refugees in the form of programs admitting refugees for permanent resettlement, response to requests for political asylum, land for refugee camps, and provision of financial and other resources for international organizations seeking to respond to refugee situations.

The demand for both international migration, temporary labor migration, refugee resettlement, and temporary asylum can be expected to continue to grow in the next decade, with those seeking to move drawing overwhelmingly from developing regions in the South. Emerging and persistent patterns of undocumented migration throughout both the developed and developing world are symptoms of the motivation of people to seek better opportunities through international migration. Matched with this demand are national doors that are gradually closing to international migrants. In the traditional receiving countries, concern over the social, economic, and political effects of immigration have moved high on the political agenda at both the national and state or provincial levels, often, but not always, with efforts to tighten migration controls, reduce immigration levels, and constrain access of immigrants to national and local social programs.

■ CONCLUSION

Population trends and patterns within countries and regions hold fundamental and inescapable implications for the full spectrum of global issues addressed in this book. While the annual rate of world population growth has been declining in recent decades, significant increases in population size, particularly in countries in the developing world, will continue into the near future. Understanding the sources of population change, specifically declines in fertility and patterns of migration, is a critical dimension of efforts to attain sustainable development and reduce poverty in regions throughout the world and to protect the global environment.

To contribute to these important discussions, this chapter has sought to introduce readers to the international and national population issues that exist and are emerging within the global arena. The discussion has been built on a foundation of basic tools for the study of population, important trends in population growth and its component demographic processes, and dominant perspectives concerning the causes and consequences of population growth. Threaded throughout the discussion has been consideration of the relationships among perspectives about population, society, and environment; models of future population growth; and the design of global and national population policies.

■ QUESTIONS

1.   Is population growth a major global problem?

2.   Do you agree more with the views of Simon or Ehrlich with reference to population growth?

3.   Should countries open their borders to refugees fleeing political persecution and/or those seeking economic opportunity?

4.   Was the Chinese government's one-child policy justified? Should governments be involved in population policy?

5.   Should the U.S. government give foreign aid to reduce world population growth? Should the aid be conditional?

■ SUGGESTED READINGS

Arizpe, Lourdes M., Priscilla Stone, and David C. Major, eds. (1994) *Population and Environment: Rethinking the Debate*. Boulder, CO: Westview Press.

Ashford, Lori S. (1995) "New Perspectives on Population: Lessons from Cairo," *Population Bulletin* 50, no. 2. Washington, DC: Population Reference Bureau.

Castles, Stephen, and Mark J. Miller (1993) *The Age of Migration: International Population in the Modern World*. New York: Guilford Press.

Commoner, Barry (1992) *Making Peace with the Planet*. New York: New Press.

Ehrlich, Paul, and Anne Ehrlich (1992) *The Population Explosion*. New York: Doubleday.

Harper, Charles L. (1995) *Environment and Society: Human Perspectives on Environmental Issues*. Upper Saddle River, NJ: Prentice Hall.

Lutz, Wolfgang, ed. (1996) *The Future Population of the World: What Can We Assume?* Laxenburg, Austria: International Institute for Applied Systems Analysis.

McFalls, Joseph Jr. (1995) "Population: A Lively Introduction," *Population Bulletin* 46, no. 2. Washington DC: Population Reference Bureau.

Moffett, George D. (1994) *Critical Masses: The Global Population Challenge*. New York: Penguin.

Myers, Norman, and Julian L. Simon (1992) *Scarcity or Abundance: A Debate on the Environment*. New York: W.W. Norton.

Women and Development

Elise Boulding

Let us consider *development* to refer to social, economic, and political structures and processes that enable all members of a society to share in opportunities for education, employment, civic participation, and social and cultural fulfillment as human beings, in the context of a fair distribution of the society's resources among all its citizenry. The United Nations (UN) bound itself in its Charter "to achieve international cooperation in solving international problems of an economic, social, cultural, or humanitarian character." In other words, the more industrialized countries of the North agreed to help the less industrialized countries of the South to reach the higher economic and social level already achieved in the North. The first thing to note about this development planning is that it has been done almost entirely by men, for men, with women and children as a residual category. How could half the human race be invisible to development planners in spite of Mao Tse-tung's well-known saying that "women hold up half the sky"? This chapter will review how this situation evolved, how it has hampered "real development," and what women—and men—are doing about it.

■ FROM PARTNERSHIP TO PATRIARCHY

In the early days of the human species, men, women, and children moved about in small hunting and gathering bands, sharing the same terrain and exploring the same spaces. Role differentiation was minimal, though child-bearing restricted women's movements somewhat, and men ranged farther in hunting prey. Even so, it is estimated that women, as roving gatherers,

157

supplied up to 80 percent of the band's diet by weight, and therefore carried an important part of the band's ecological knowledge in their heads. Based on observation of small hunting-gathering tribes found in Africa, Australia, and South America today, it would seem that old women as well as old men took the role of tribal elders and carried out rituals important to the social life of the band. About 12,000 B.C.E., a combination of events brought about a major change in the human condition; improved hunting techniques resulted in a dwindling supply of animals, and women discovered from their plant-gathering activities that seeds spilled by chance near last year's campsite would sprout into wheat the following year, creating a convenient nearby source of food. This resulted in the deliberate planting of seeds to grow food, and agriculture came into being. This changed everything. Since working with seeds was women's work, women became the farmers, and men went farther and farther afield in search of scarce game, thus discovering exciting new terrains that women knew nothing about. The shared-experience worlds of women and men were now differentiated. Women knew about everything that was close to home, and men were exploring the world "out there."

While a lot has happened to humankind since 12,000 B.C.E., bearing children, growing and processing food, and feeding families (also creating shelters and clothing) have continued to be women's work for most of the world's population. From the earliest farming settlements to villages to towns to cities and civilizations took another 8,000 to 10,000 years, but cities never returned to women that freedom of movement they had had as gatherers. Rather, they heralded the rise of the rule of men, or patriarchy, to replace an earlier partnership. What cities and civilization brought were concentrations of wealth and power—and houses that enclosed women and shut them off from the outside world. Only poor women were free to scurry about the streets to provide services for the rich. With industrialization, and major population movements to the cities where the factories were located, there were fewer and fewer women enjoying the relative freedom of farming and craft work, which families often carried out as a family team in the period known as the "Middle Ages." Now they were either shut up in the home or shut up in factories.

While I have been describing major trends for women, reality for individual women was much more complex, and in every age women found ways to be creative and to improve the circumstances in which they lived. While most of women's creativity was invisible, we do find women— queens and saints and philosophers and poets—in the history books. We also find precursors of the contemporary women's movement from the 1500s through the 1700s. However, by the nineteenth century, the pressures of urbanization and industrialization in Euro–North America gave rise to a new social sector of educated middle-class women with free time and a growing awareness of the world around them, including awareness of women migrants from rural areas, chained with their children to factory and slum. This produced a small group of women radicals and revolutionaries and a larger group of liberal reformers and concerned traditionalists who translated their sense of family responsibility into responsibility for the community. These women quickly discovered, when they intruded into the political arena, that as women they had no civic or legal identity, no political rights, and no economic power. The realization that they needed civic rights to get on with reform led to an exciting century and a half of mobilization of women in the public arena. Their concerns about economic conditions for the poor soon spilled over to concern about the frequent wars that rolled over Europe, which they saw as directly related to poverty and suffering. The suffrage movement came into being because women realized that they needed political power in order to be able to eradicate the social evils they saw. Everywhere, women came up against the rule of men and sought partnership with them rather than domination by them.

The patriarchal model pervades society but begins in the family with the rule of the male head of household over wife or wives and children; it has served as a template for all other social institutions, including education, economic life, civic and cultural life, and governance and defense of the state. Because so many generations of humans have been socialized into the patriarchal model, the struggle to replace it with partnership between women and men will be a long one; and it has barely begun. Now, however, the survival of the planet is at stake. Women's knowledge of their social and physical environment, of human needs, of how children learn, of how conflicts can be managed without violence and values protected without war—as well as their skill in managing households with scarce resources—are urgently needed wherever planning takes place and policies are made. Yet, these are precisely the places where women are absent. Now we will turn to the story of how women are beginning to sow the seeds of transformation of a system based on domination over the very ones who hold the skills to shape a better life for all. First, we look at the United Nations and its role in this process.

■ THE UNITED NATIONS: THE DEVELOPMENT DECADES

As Western colonial empires began to dissolve in the UN's first decade, it became clear that the ever more numerous new, poor nations that came to be called the "third world" were going to need a lot of assistance from the founding states of the UN in order to work their way out of poverty. The popular "trickle-down" theory led to the encouragement of capital-intensive heavy industry in the tiny modern enclaves of poor new states at the expense of more labor-intensive light industries that might employ the un-employed and the unschooled. The only aid given to agriculture was in the form of capital-intensive equipment that required large acreages and pushed subsistence farmers onto ever more marginal lands with rapidly de-teriorating soils. This approach, labeled economic dualism, leaves the bulk of a country's population without the skills and tools needed to be more productive. By the end of the 1950s, it was evident that "international co-operation" was not helping poorer countries at all. Then the UN General Assembly declared the 1960s a Development Decade, a time of catching up for all the societies left behind in the twentieth-century march of progress. Yet, the policies based on economic dualism remained un-changed, in spite of valiant efforts by the UN Development Programme (UNDP), established in 1965, to undertake a more diversified approach. By the end of that decade, gross national product (GNP) growth rates were negative in a number of third world countries, even in some that had "good" development prospects.

A second Development Decade was launched in 1970 to try to do what the first had failed to do, but by then it had become clear to many third world countries that the development strategies recommended by "first world" countries were not working. They formed their own Group of 77 (eventually including 118 states) to promote a Program of Action for the establishment of a New International Economic Order. This called for more aid from countries of the North, debt moratoriums, and other strategies for preventing the economic gap between the North and South from continuing to increase. A monitoring system for the conduct of multinational corporations was also demanded. The North had no serious interest in acting on these proposals, and the poor nations kept getting poorer. In 1980, the UN launched yet another Development Decade, still following failed policies. One problem faced was the World Bank's struc-tural adjustment program, which compelled countries of the South to focus on producing cash crops for the international market to reduce their indebtedness, and to spend less money on human services. This intensi-fied the dualism between a low-productivity agricultural sector and a high-productivity agribusiness and industrial sector. People in the South went hungry and without schooling while food was exported to the North (since more money could be made), and countries sank still deeper into poverty.

The goal of all this Development Decade activity, planned and carried out by men, was to increase the productivity of the male worker. But one quiet day in the late 1960s, a Danish woman economist, Ester Boserup, de-cided to study what was actually going on in that intractable subsistence sector of agriculture. She was able to point out that the majority of the food producers were women, not men, but all agricultural aid, including credit and tools, was given to men. This aid not only failed to reach the women but encouraged men to grow cash crops (which needed irrigation) for export. This added to the women's work load, as the men demanded that their wives tend the new fields in addition to the family plot that fed the family and provided a modest surplus for the local market.

Now a whole new picture began to emerge. Boserup focused her studies on Africa, which was bearing the cruelest load of suffering in the food supply crisis. She found that roughly 75 percent of food producers were women, often either sole heads of households or with migrant husbands living and working elsewhere. Their working hours were on the average fifteen hours a day. They worked with babies on their back, and their only helpers were the children they bore. These women farmers never received advice on improved methods of growing food, tools to replace their digging sticks, wells to make water more available, credit to aid them during the years of bad crops, or aid in marketing. Long hours of walking to procure water and to get to market (with their produce carried on their heads), in addition to the hard work in the fields, made their lives a heavy struggle.

Yet, emphasizing the hardship women experience should not obscure the importance of women's special knowledge stock, which is basic to community survival in any country, in both rural and urban settings. In a future world with more balanced partnering between women and men, knowledge will not be so gender linked; for now, it is imperative that it be recorded, assessed, and used. Women's special knowledge stock relates particularly to the six following areas: (1) life-span health maintenance, including care of children and the elderly; (2) food production, storage, and short- and long-term processing; (3) maintenance and utilization of water and fuel resources; (4) production of household equipment, often including housing construction; (5) maintenance of interhousehold barter systems; and (6) maintenance of kin networks and ceremonials for handling regularly recurring major family events as well as crisis situations. While only (2) relates directly to food, all six factors contribute to the adequate nutrition of a community; (6) is particularly important in ensuring food sharing over great distances in times of food shortages and famine. Unfortunately, the extended-kin/ceremonial complex is one of the first resources to be destroyed with modernization.

The story of the UN's Third Development Decade could have been different if policy planners had had this information at their fingertips. But now there was a women's movement ready to hear what Ester Boserup and a growing group of women development professionals had to say about women's roles in the development process. The UN came to play an important part in giving them a platform.

■ THE UNITED NATIONS: THE WOMEN'S DECADES

In 1963, the UN General Assembly proclaimed 1965 as International Co-operation Year (ICY). Just two years before, the Women's Strike for Peace movement had spread from the United States to other continents, so when word got out that 1965 would be a UN International Cooperation Year, an international women's network quickly sprang into being to organize projects for the year in many countries.

Between the women's inexperience and the international bureaucracy's inertia, the ICY fell short of its sponsors' ambitious hopes; but for the international community of women, it proved to be the start of a still unfolding chain of events. Women formed traveling teams to every continent and began seeing with their own eyes in the rural areas of Africa, Asia, and Latin America what Ester Boserup was documenting as an economist. They began to see women as the workers of the world, though invisible to census enumerators because much of their productive labor takes place in the world of the home: the kitchen, the kitchen garden, and the nursery. "Employed persons" were supposed to have identifiable outside jobs and wages. Unwaged labor and labor in what is called the informal economy (which is not reported in the economy's record books) were concepts not taken seriously by economists and statisticians.

This did not stop the women who traveled together in the ICY teams. As they began to see homemakers as workers, they redefined their own role. They liked to be called "housekeepers of the world." "International housekeeping" brought them in 1972 to the UN Conference on the Human Environment in Stockholm and in 1974 to the World Population Conference in Bucharest and the World Food Conference in Rome. Each time, they came in larger numbers and with better documentation on how the conference subject was relevant to women. They also became increasingly aware of how blind most of their male colleagues were to the importance of women's roles in economic production and social welfare. Since governments appointed few women to international conferences, the knowledge and wisdom of this growing group of observers went unheard. They were outsiders, petitioners, protesters.

Nevertheless, the UN had actually established, in its Economic and Social Council, a UN Commission on the Status of Women as early as 1946, and by 1948 the UN Declaration on Human Rights included gender as a category. In fact, over the decades the UN adopted twenty-one conventions on the rights of women, although they remained largely paper rights because member states did not provide for the implementation of the rights. It was still a man's world. But women were learning strategies for working in it. The Commission on the Status of Women in particular was building up experience in working within the UN system and began to assert itself. Its members saw to it that a phrase about the importance of integrating women into development was included in the development program for the Second Development Decade. While ignored by decision-makers, the phrase immediately reverberated in the growing international women's movement. Furthermore, in 1972, Helvi Sipila of Finland was appointed the first woman UN assistant secretary-general, for humanitarian and social affairs. Empowered by Sipila's support, the Commission on the Status of Women worked closely with the older women's organizations and the newer networks to create a women's agenda for the UN.

The first real breakthrough came when the UN General Assembly declared that 1975 would be International Women's Year. The Women's World Plan of Action (UN 1976), drafted for the 1975 assembly of women in Mexico City, defined *status* in terms of the degree of control women had over their conditions of life. This became a key theme that continues to this day in the women's movement. The International Women's Year became the UN International Decade for Women (or Women's Decade), with follow-up meetings in 1980 in Copenhagen, in 1985 in Nairobi, and in 1995 in Beijing. These UN-sponsored world women's conferences have represented a growing voice for women as participants and coshapers of the world in which they live. The trio of themes for each of these gatherings, which swelled from 6,000 women in Mexico City to 14,000 women in Nairobi to 50,000 in Beijing, has been: equality, development, and peace. The interrelationship of these three concepts in the lives of women, and in the life of every society, represents an important breakthrough in the conceptualization of development.

The guidelines for national action laid down for the Women's Decade are as relevant today as they were twenty years ago:

- Involving women in the strengthening of international security and peace through participation at all relevant levels in national, intergovernmental, and UN bodies
- Furthering the political participation of women in national societies at every level
- Strengthening educational and training programs for women
- Integrating women workers into the labor force of every country at every level, according to accepted international standards
- Distributing more equitably health and nutrition services to take account of the responsibilities of women everywhere for the health and feeding of their families
- Increasing governmental assistance for the family unit
- Involving women directly, as the primary producers of population, in the development of population programs and other programs affecting the quality of life of individuals of all ages, in family

groups and outside them, including housing and social services of every kind

Immediate outcomes of the 1975 conference included the establishment of the UN International Research and Training Institute for the Advancement of Women (INSTRAW), the UN Development Fund for Women (a Voluntary Fund), and a modest working relationship with the UN Development Programme. With few resources and very small staffs, the two new UN bodies have nevertheless played an important part in the gradual recognition of women as actors, not only as subjects needing protection.

In short, women were no longer to be treated as invisible or as subject to patriarchal rule. They are persons, citizens, and actors on the local and global scene. The importance of the role of the UN in providing an international platform for women to be seen, heard, and listened to with respect cannot be underestimated—even though the UN itself still has a long way to go in making senior posts available to women. Now the concepts are *there*, in public international discourse, and the UN and its member states cannot completely ignore them, however much they drag their feet in applying them. This change in visibility could not have happened without a strong involvement of women from every continent. How did the invisible become visible so quickly?

■ WOMEN'S NETWORKS REDEFINE THE MEANING OF DEVELOPMENT

As mentioned earlier, women were already active internationally on behalf of the oppressed poor by the middle of the nineteenth century. They came to know each other across continents by meeting at the great world's fairs—London in 1851, Paris in 1855 and 1867, and Chicago in 1893. These were women of the upper middle classes who could travel, of course. Many innovations in education and welfare services and home services for working women and children, prisoners, and migrants emerged from those meetings. By 1930, there were thirty-one international nongovernmental women's organizations (INGOs or NGOs) and more of them had working-class members. However, it was not until the United Nations brought grassroots women from each continent in touch with each other that there emerged a deeper understanding of the problems that women faced in countries formerly colonized by the West. While every country experienced gender-based dualism (women's only and men's only jobs), only in the third world, which happens to have two-thirds of the world's population (and which will be referred to from now on as the two-thirds world), was gender-based dualism linked to an economic dualism that trapped women in the subsistence sector. In other words, women are left without the skills or opportunity for economic improvement.

While the cultural specifics of the situation of women differed between countries and continents, they all had a common experience base linked to the fact of bearing and rearing children and having responsibility for the nourishment, health, and well-being of family members. Thus, while men measured development in economic terms of rates of growth of the gross national product, women's thinking was in terms of human and social development. What made life better for individual human beings and what made societies more humane and joyful to live in? These were the questions women were asking as they formed numbers of new women's INGOs committed to human rights, development, and an end to violence. The same questions were being asked by the new women's networks that began forming during preparations for the first International Women's Year.

The first of the new networks was formed at the International Tribunal on Crimes Against Women, which met in Brussels in 1974. From that tribunal we get the first major international statement about patriarchal power as violence against women per se, apart from specific abusive acts. Participants joined to establish the International Feminist Network after listening to a horrifying array of types of violence experienced by women from all classes on all continents at the hands of men in their families, communities, and places of work. Today we are aware that this violence extends to the horror of wartime rape, an old but increasingly visible crime against women wherever the violence of war takes place.

As awareness of the obstacles to achieving the good life not only for women but for society as a whole increased, new transitional networks developed and multiplied from the time of the tribunal to the present. Thus, internationally women are strongly aware that obstacles persist to their full partnership in the development process and that violence and other human rights abuses and war itself constitute a significant part of those obstacles.

Since being able to work with men as equals is a precondition for women's voices to be heard, equality has been a priority goal at least since the beginning of the suffrage movement. But gaining the vote, with some associated legal and property rights, did not improve women's economic situation, nor did it bring sexual discrimination to an end or move women toward policymaking positions in society. Therefore, one part of the equal rights movement has dealt in painstaking detail with legal rights in the area of family, employment, rights to education and training, and equal access to opportunity in general. This is important for women everywhere, but especially in the poorest countries, where men's literacy rate is barely at 40 percent and women's literacy rate is only a third or a half of the men's rate. As opportunities for advanced study begin to become more available to women, the increase in women graduating with degrees in law, engineering, and the sciences, both physical and social, means that the sisterhood for social change on each continent has been able to draw on a very high level of competence and expertise in their work for equality.

Another part of the equal rights movement has taken a more direct political track: women have sought to be involved in the lawmaking process as elected officials. Not surprisingly, the first visible change has been at the grassroots. During the Women's Decade it was reported that, starting from zero before women could vote, women's representation on local government councils had risen to 11 percent worldwide, with countries at the low end ranging from 0 to 6.4 percent, at the high end up to 49 percent. At the national level, 7.3 percent of women were elected to national parliaments, with countries at the low end ranging from 0 to 5 percent, at the high end up to 33 percent. At the global level, 4.3 percent of women served as appointed officials in the diplomatic sphere, with countries at the low level ranging from 0 to 3 percent, at the highest level up to 30 percent. Women's political participation may be thought of as a pyramid, with the bulk of women found at the bottom, at the local level, and the fewest at the top, at the diplomatic level. In general, there is a "tipping point" phenomenon for women in elected or appointed office. When there are a very few, "token" women, they tend to confine themselves to what are thought of as traditional women's issues, notably the well-being of families and children. As women grow in numbers and self-confidence, however, they are empowered to address broader social system issues that affect all sectors of society.

As professional skills are bringing women into the legislative process, their know-how is also becoming available to development specialists, including the language of "development alternatives" and "another development." Instead of high-tech industrial enclaves and agribusinesses that rob the poor of their intensive hand labor, alternative development means investing in tools and water supplies for farmers and craft workers; making credit available for small improvements, including equipment for small local factories; and building local roads, schools, and community centers that can serve many needs, including day care for small children and health services.

While women have been leaders in all these efforts to achieve human and social development, not simply economic development, they have certainly not been alone. Ever since E. F. Schumacher (1993) wrote about "economics as if peopled mattered," there have been creative and humanistically oriented male development professionals who have worked both alone and together with women to further these broader goals. They have played their role also in helping to strengthen the women's networks.

Protecting the environment from multinationals that are logging whole forests and leaving fragile soils to erode, that are undertaking mining operations destructive of local farmland and waterways, that are building dams that destroy huge acreages and cause an increase in homelessness and unemployment—these activities also become an important agenda for women's networks. They are proud to bear the label *ecofeminists*. (Note: It was a woman, Ellen Swallow, who coined the term *ecology* in 1892, and it was a woman, Rachel Carson, who gave the signal for the birth of the modern ecology movement with her *Silent Spring*, appearing in 1962.)

■ FROM PATRIARCHY TO PARTNERSHIP

We are midstream in a long, slow process of transformation. As described more fully at the end of *The Underside of History: A View of Women Through Time* (Boulding 1992), in one sense the institutions of society continue to be stacked against women. There are strong expectations of subservient behavior on the part of women, reinforced by upbringing, teachings of church and school, continuing inequality under the law, un-derrepresentation in government, and media portrayal of women as consumer queens. However, we must not underestimate the fact that there has always been a women's culture inside the patriarchal culture that modifies and changes patriarchal institutions. We can therefore think of women as shapers, not only victims, of society. However, individual initiative alone can be weak and ineffective. The support system for women found in women's organizations and networks acts as a great multiplier of individual effort. Below are three case studies that illustrate many of the ideas discussed in the preceding paragraphs.

■ SOME COUNTRY EXAMPLES OF THE WOMEN-DEVELOPMENT RELATIONSHIP

■ India

In a remote village in Northern India called Reni, nestled in the Garwhal mountains of the Himalayan range, a group of women and children were doing an unusual thing. Hugging trees. They were resorting to Chipko, which means to hug, to prevent trees from being felled. With their arms wrapped around the trees, the women and children cried: "The forest is our mother's home, we will defend it with all our might." The women and children stopped about 70 lumberjacks from the forest contractors who were to fell the trees. (Anand 1983)

With country after country selling off its forests to multinationals, not only displacing forest-dwelling people and removing their source of sustenance but destroying the very lungs of the earth and endangering the entire ecosphere, the Chipko movement has spread from village to village and country to country. It is mostly women's work to gather the rich supplies of forest plants, roots, berries, and tree fruits to feed their families and sell surpluses in the market for modest cash returns. It is their careful pruning of the trees for firewood that provides the only fuel for cooking. Therefore, women know they must be the protectors of the forests. They also know that trees help store water in the soil, that both water and the quality of soil itself will be lost when land is stripped of tree cover. And they know that financial resources needed for local development are in danger of being siphoned off to create roads and other infrastructure for an export-dominated economy, leaving the village poorer than ever. Already, women in the Gharwal area had to walk 5 to 10 miles a day to collect firewood and fodder for cattle, bent double under heavy baskets.

In the local village councils, the leaders and decisionmakers are men, and these initiatives by women have startled them. Some villages are proud of their women and support them. Others know they are doing what is needed but feel that it is men who should be leading the way (only they do not). While there have been some sharp disagreements between women and men at the village level, the women have been able to sustain the courage of their convictions, supported by educated urban women like Vandana Shiva, director of the Research Foundation for Science, Technology, and Natural Resource Policy in Dehradun, India. Shiva has been an articulate voice linking care for the environment to a model of alternative development. She insists on the importance of building on grassroots capabilities that will remove the gulf between rich and poor. The village women now know they have a voice, and that voice is beginning to be listened to locally, nationally, and internationally.

▪ Kenya

The Green Belt movement for reforestation is responding to a different phase of the same environmental crisis that the Chipko movement is responding to. The trees of the Rift Valley in Kenya are already gone, and the land is undergoing soil erosion and desertification on a large scale. The Green Belt movement brings the women of local villages into tree-planting activities in order to rebuild seriously degraded soils, and also to provide the foods and fuelwood that a well-cared-for forest can provide. The plan is very simple. The women of each village plant a community woodland of at least 1,000 trees, a green belt. They prepare the ground, dig holes, provide manure, and then help take care of the young trees.

The inspiration for this movement came from a Kenyan woman biologist, Professor Wangari Maathai, who saw that the natural ecosystems of her country were being destroyed. As she has pointed out, the resulting poverty was something men could run away from, into the cities, but the women and children remain behind in the rural areas, hard put to provide food and water for hungry families. Working with the National Council of

Women of Kenya, Professor Maathai went from village to village to persuade women to join the Green Belt movement. Green Belt by now has involved more than 80,000 women and a half million school children in establishing over a thousand local tree nurseries and planting more than 10 million trees in community woodlands, with a tree survival rate of 70 to 80 percent. This revival of local ecosystems has made fuel available, and family kitchen gardens are once more viable. Furthermore, women have been empowered to develop more ways of processing and storing foodstuffs that were formerly vulnerable to spoilage.

This movement, like the Chipko movement, has spread to other countries. In Kenya, as in India, educated urban women were in close enough touch with their village sisters to be aware of their needs. The basic development issue at stake, says Professor Maathai, is food security, without which no country can have any meaningful development. Women helped introduce the food security concept to development professionals.

Bangladesh

Long before the days of colonialism, the traditional practice of forming local credit associations had tided many women over hard times. The new situation of a growing cash economy in both rural and urban areas greatly increased the need for credit, but credit is rarely available to women through local banks. In Bangladesh, the Grameen Bank was started specifically for the poorest sector of the population and now has thousands of borrowers' groups throughout the country, three-fourths of them women. A network of roughly 8,000 "bankers on bicycle," each trained by the Grameen organization, covers many areas of Bangladesh, and the Grameen principle has now spread to other Asian countries, to Africa, and to Latin America. Further loans to a borrowers' group depend on repayment of previous loans, and repayment rates average 98 percent of all loans, with women having consistently higher repayment rates than men. Once a women's group has successfully completed its first round of loans for land for farming, for livestock, or for tools, they often expand their activities to building schools, clinics, and needed local production facilities. Studies of Grameen find that the incomes of borrowers' groups in Grameen Bank villages are 43 percent higher than that of borrowers in non-Grameen villages. This is a tribute both to the Grameen method and to the business acumen of women borrower groups.

ASSESSING WHERE WE ARE

We have followed the development decades from the perspective of women and noted the absence of women from the planning process. Male planners' lack of knowledge of the importance of women's agricultural labor, and of their economic and social contributions to a national standard of living, led development professionals and the UN development institutions, including the World Bank, to concentrate on a capital-intensive type of development. This included agricultural development that remained in special enclaves. Nothing "trickled down" to poor farmers, mostly women, and the countries of the South grew poorer as resources were diverted from the areas of greatest need.

We have also seen how the international women's movement, with support from the UN, has set about to address the ignorance of male planners about women's work. At the same time, women's efforts to promote equality of participation in development, starting with the Women's Decade, has led to a growing awareness that overcoming poverty and improving the quality of life for all requires a different relationship between women and men, moving from patriarchy to partnership. When women and men can share their experience worlds, resources of the World Bank and member states will be better used at local levels.

This is already happening. The World Bank has started modifying its structural adjustment programs. In the spring of 1997, a noteworthy global campaign was launched by INGOs and private sector antipoverty groups, with the backing of the World Bank, the UN Development Programme, national leaders, international aid agencies, foundations, and corporations, to make small loans to nearly 100 million poor people to finance small-scale farming and trade, with a special focus on working with local women farmers and entrepreneurs. Women's World Banking has been one of the leading promoters of this campaign, which will result in a tremendous multiplication of Grameen Bank–type projects. This represents an important step toward understanding development as human and social development, opening up more possibilities for diverse approaches to human betterment—recognizing, as Schumacher said a long time ago, that "small is beautiful." Most of all, it points to a growing partnership between the women and men who will be working for that more diversified, more earth-loving, more local and yet more connected world of the future.

■ QUESTIONS

1. After reading this chapter, what does development mean to you?

2. Why have the international women's conferences from 1975 to 1995 continued with the same three themes: equality, development, and peace? Do you agree that these themes belong together?

3. Why does the author prefer the term *two-thirds world* to the commonly used *third world?*

4. The author focuses primarily on problems of development in the two-thirds world. Do you think the United States has development problems?

5. What does the phrase *from patriarchy to partnership* mean? What would it mean for U.S. society?

■ SUGGESTED READINGS

Boserup, Ester (1970) *Women's Role in Economic Development*. New York: St. Martin's Press.

Boulding, Elise (1992) *The Underside of History: A View of Women Through Time*. Revised edition. Newbury Park, CA: Sage Publications. (See Vol. 1, Ch. 4, "From Gatherers to Planters"; Vol. 2, Ch. 5, "The Journey from the Underside: Women's Movements Enter Public Spaces.")

Caldecott, Leonia, and Stephanie Leland, eds. (1983) *Reclaim the Earth: Women Speak Out for Life on Earth*. London: Women's Press.

Fisher, Julie (1993) *The Road from Rio: Sustainable Development and the Nongovernmental Movement in the Third World*. Westport, CT: Praeger.

Masini, Eleanora, and Susan Stratigos, eds. (1991) *Women, Households and Change*. Tokyo: UN University Press.

Shiva, Vandana (1989) *Staying Alive: Women, Ecology and Development*. London: Zed Books.

Turpin, Jennifer, and Lois Ann Lorentzen (1996) *The Gendered New World Order: Militarism, Development and the Environment*. New York: Routledge.

United Nations (1995) *Fourth World Conference on Women*. A/Conf. 177/20, October 17. New York: United Nations.

Waring, Marilyn (1988) *If Women Counted: A New Feminist Economics*. San Francisco: Harper & Row.

Children

George Kent

Worldwide many children live in wretched conditions, suffering from malnutrition and disease, laboring in abusive work situations, and suffering exploitation of the most grotesque forms. The gravest problems of children are found in the third world, but even in industrialized nations many children are severely disadvantaged. In the United States, for example, fully one-fifth of the nation's children live below the official poverty line. My purpose here is to show that the situation of children should be understood not merely as a series of unconnected localized and private problems, but as systemic problems of public policy requiring attention at the highest levels of national and international governance.

Increasing attention by policymakers to the problems of children have resulted in some real progress in improving the quality of children's lives. The advances are documented every year in reports of the United Nations Children's Fund (UNICEF): *The State of the World's Children* and *The Progress of Nations*. In 1993, for example, *The Progress of Nations* reported:

> In little more than one generation, average real incomes have more than doubled; child death rates have been more than halved; malnutrition rates have been reduced by about 30%; life expectancy has increased by about a third; the proportion of children enrolled in primary school has risen from less than half to more than three quarters; and the percentage of rural families with access to safe water has risen from less than 10% to more than 60%. (UNICEF 1993a: 4)

However, satisfaction with such successes must be tempered with appreciation of the great distance still to be traveled if all children are to live a life of decency. Perhaps the clearest lesson learned in recent years is that significant gains in children's well-being do not result from economic growth alone. They also require progressive social policy based on a sustained commitment to improvements in the well-being of the poor in general and children in particular.

The following four sections provide an overview of the situation of children with regard to *child labor*, *child prostitution*, *armed conflict*, and *malnutrition*. A section on *child mortality* then shows that such pressures on children result in massive mortality, making even armed conflict look relatively unimportant by comparison.

While many different kinds of programs have been developed over the years to address the concerns of children, most have been inadequate to the task. There is now new hope in the rapidly advancing recognition of *children's rights*, based on the acknowledgment that every single child has the right to live in dignity. The legal obligation for the fulfillment of children's rights falls primarily on national governments, but for these large-scale global issues there are *international obligations* as well. The last section of this chapter focuses on these obligations.

■ CHILD LABOR

Children work all over the world, in rich as well as poor countries. They do chores for their families, and many go out to fields and factories to earn modest amounts of money. Children's work can be an important part of their education, and it can make an important contribution to their own and their families' sustenance. There can be no quarrel with that. The concern here, however, is with child *labor*. Child labor can be defined as children working in conditions that are excessively abusive and exploitative. It is not clear where exactly the boundary line between acceptable children's work and unacceptable child labor should be located, but there are many situations in which there can be no doubt that the line has been crossed. Abdelwahab Bouhdiba, in a study on *child labor*, offered many illustrations:

> Thousands of girls between the ages of 12 and 15 work in the small industrial enterprises at Kao-hsiung in southern Taiwan. . . . Some children [in Colombia] are employed 280 metres underground in mines at the bottom of shafts and in tunnels excavated in the rock. . . . Most carpet-makers [in Morocco] employ children between the ages of 8 and 12, who often work as many as 72 hours a week. . . . [In Pakistan] slave traffickers buy children for 1,600 rupees from abductors. They cripple or blind the weakest, whom they sell to beggar masters. . . . One million Mexican children are employed as seasonal workers in the United States. (Bouhdiba 1982: 2, 3, 11, 20)

Many children are caught up in the bonded labor system, especially in South Asia and Latin America. In the succinct explanation provided by the International Labour Organization's *World Labour Report 1993*:

> The employer typically entraps a "bonded" labourer by offering an advance which she or he has to pay off from future earnings. But since the employer generally pays very low wages, may charge the worker for tools or accommodation, and will often levy fines for unsatisfactory work, the debt can never be repaid; indeed it commonly increases. Even the death of the original debtor offers no escape; the employer may insist that the debt be passed from parent to child, or grandchild. Cases have been found of people slaving to pay off debts eight generations old. (ILO 1993: 11)

In Pakistan, an estimated 20 million people work as bonded laborers, 7.5 million of them children. The carpet industry alone has perhaps 500,000 bonded child workers. Afghan refugees in Pakistan, and their children, are now included in Pakistan's pool of bonded laborers.

Children work in rich countries as well. In the United States, for example, in 1988 about 28 percent of all fifteen-year-olds were working. The United States General Accounting Office (USGAO 1991) found that of the employed fifteen-year-olds, about 18 percent worked in violation of federal child labor regulations governing maximum hours or minimum ages for employment in certain occupations. Many working teenagers are injured on the job. Enforcement of child labor laws has been weak in many states, apparently due to the greater concern with protecting the interests of employers.

Paradoxically, the acceptance of child labor tends to be higher where there arc higher surpluses of adult labor. The addition of children to the labor force helps to bring down wage rates, which in turn makes it more necessary to have all family members employed. The widespread employment of children keeps them out of school and thus prevents the buildup of human capital that is required if poor nations are to develop.

In sum, millions of children work, many under grossly exploitative conditions. The ILO has estimated that in the year 2000 there will still be at least 37 million working children under the age of fifteen (ILO 1993).

■ CHILD PROSTITUTION

Child prostitution refers to situations in which children engage in regularized sexual activity for material benefits for themselves or others. These are institutionalized arrangements—sustained, patterned social structures—in which children are used sexually for profit. Child prostitution is an extreme form of sexual abuse of children and an especially intense form of exploitative child labor. Most prostitution is exploitative, but for mature men and women there may be some element of volition, some consent. The assumption here is that young children do not have the capacity to give valid, informed consent on such matters.

Child prostitution is widespread. It has been estimated that about 5,000 boys and 3,000 girls below the age of eighteen are involved in prostitution in Paris. The Ministry of Social Services and Development in the Philippines has acknowledged that child prostitution rivals begging as the major occupation of the 50,000 to 75,000 street children who roam metropolitan Manila. The number of underage prostitutes in Bangkok numbers at least in the tens of thousands. In India the number is surely over 100,000. It has been estimated that there are about 600,000 child prostitutes in Brazil. The number of child prostitutes worldwide is probably well over a million.

In some places, such as India and Thailand, child prostitution was deeply ingrained as part of the culture well before foreign soldiers or tourists appeared in large numbers. There are many local customers. Some Japanese and other tourists may use the child prostitutes in the "tea houses" in the Yaowarat district of Bangkok, but traditionally most of their customers have been locals, especially local Chinese. Similarly, in the sex trade near the U.S. military bases in the Philippines before they closed down, more than half the customers were local people. There is big money associated with the foreign trade, but there are bigger numbers in the local trade.

■ ARMED CONFLICT

Armed conflicts hurt children in many ways. Wars kill and maim children through their direct violence. Children are killed in attacks on civilian populations, as in Hiroshima and Nagasaki. In Nicaragua, many children were maimed or killed by mines. The wars in Afghanistan in the 1980s and in Bosnia in 1993 were especially lethal to children. Wars now kill more civilians than soldiers, and many of these civilians are children. Children have been counted among the casualties of warfare at a steadily increasing rate over the past century. Historically, conflicts involving set-piece battles in war zones away from major population centers killed very few children. However, wars are changing form, moving out of the classic theaters of combat and into residential areas where civilians are more exposed.

There is also a great deal of violence against children in repressive conditions short of active warfare. Thousands of street children have been killed with impunity by death squads in Latin American countries.

Children are frequently hurt in the aftermath of warfare by leftover mines. The International Committee of the Red Cross has estimated that "using current mine-clearing techniques, it would take 4,300 years to render only twenty per cent of Afghan territory safe" (ICRC 1993: 471).

Often children are pressed to participate in armed combat as child soldiers, harming them both physically and psychologically. Children can be the agents as well as the victims of violence. Increasingly, older children (ten to eighteen years old) are engaged not simply as innocent bystanders but as active participants in warfare.

Dorothea Woods, associated with the Quaker United Nations Office in Geneva, has dedicated herself to chronicling the plight of child soldiers in a monthly survey of the world's press entitled *Children Bearing Military Arms*. In the January and February 1997 editions, for example, she cites these cases:

- *Afghanistan:* "Hundreds of thousands of youth . . . were being raised to hate and fight a 'holy war'. . . . Many of those children are now with the Taliban army."
- *Chechnya:* "Government security forces have often detained young males between the ages of 14 and 18 as potential combatants in order to prevent them from joining the rebel forces."
- *Liberia:* "Because of the socio-economic crisis a part of the youth population is inclined to join one of the factions. The possession of a Kalachnikov gives the means to live by pillage and racketeering if necessary." "Various estimates have put the total number of Liberian soldiers below the age of 15 at around 6,000."
- *Sierra Leone:* "After the outbreak of the civil war in 1991 some five thousand youngsters joined either the governmental army or the rebel Revolutionary United Front."
- *Burma/Myanmar:* "A Shan boy . . . had been a porter-slave to carry heavy things to the place of fighting. . . . He fell down and was kicked by a Burmese soldier . . . until his leg broke like a stick in three places."
- *Guatemala:* "Forcing the under 18's from the indigenous communities to enroll in the army practically severs and destroys the future of these communities."
- *Mozambique:* "For the 10,000 children who took part in the civil war, the war is not over; it has been replaced by a multitude of small wars in their heads."
- *Uganda:* "Some 3,000 children have been kidnapped in the northern part of Uganda in the last four years according to UNICEF. The guerrillas who took these children have enrolled the boys in their army and have forced the girls to 'marry' the soldiers."

Wars sometimes harm children indirectly, through their interference with normal patterns of food supply and health care. Many children died of starvation during the wars under the Lon Nol and Pol Pot regimes in Kampuchea (Cambodia) in the 1970s. In 1980–1986 in Angola and Mozambique, about half a million more children under five died than would have died in the absence of warfare. In 1986 alone, 84,000 child deaths in Mozambique were attributed to the war and destabilization. The high mortality rates in Angola and Mozambique were due not only to South Africa's destabilization efforts but also to their civil wars. The famines in Ethiopia in the mid-1980s and again later in that decade would not have been so devastating had it not been for the civil wars involving Tigre, Eritrea, and other provinces of Ethiopia. Civil war has also helped to create and sustain famine in Sudan.

The interference with food supplies and health services is often an unintended by-product of warfare, but in many cases it has been very deliberate. In some cases, the disruption of the infrastructure can have deadly effects well beyond the conclusion of the war. It has been estimated that more deaths resulting from the Gulf War occurred after the war than during it.

■ MALNUTRITION

It is widely accepted that if a child's weight is more than two standard deviations below the reference for his or her age (about 80 percent of the reference weight), that child should be described as malnourished. In 1995, almost 160 million children under five years of age were seriously underweight. Contrary to the common belief that the problem is most widespread in Africa, there are far more malnourished children in Asia than in Africa. More than half the developing world's seriously underweight children are in South Asia (UNACC/SCN 1997).

During the period 1985–1995, the proportion of children in developing countries who were seriously underweight declined from about 34 percent to 29 percent. However, while the percentage declined, because of population growth, the total number remained at around 160 million throughout that period (UNACC/SCN 1997).

There has been good progress on many issues of concern to children, but in 1996 the Secretary-General of the United Nations reported that there has been little progress in reducing child malnutrition. In sub-Saharan Africa and South Asia, the number of malnourished children is actually rising. Almost a third of all children under the age of five in developing countries are malnourished, and malnutrition still contributes to more than half the deaths of young children in these countries (UNICEF 1996a).

■ MORTALITY

Nothing conveys the plight of children worldwide as clearly as their massive mortality rates. Estimates of the number of under-five deaths for selected years are shown in Table 11.1.

The number of children dying each year has been declining, but the numbers are still enormous. Children's deaths account for about one-third of all deaths worldwide. In northern Europe or the United States, children account for only 2 to 3 percent of all deaths. In many less developed countries, more than half the deaths are deaths of children, which means there are more deaths of young people than of old people. The median age at death in 1990 was five or lower in Angola, Burkina Faso, Ethiopia, Guinea, Malawi, Mali, Mozambique, Niger, Rwanda, Sierra Leone, Somalia, Tanzania, and Uganda. This means that in these thirteen countries, at least half the deaths were of children under five. In the United States, the median age at death in 1990 was seventy-six, and in the best cases—Japan, Norway, Sweden, and Switzerland—it was seventy-eight (Kent 1995).

The number of children who die each year can be made more meaningful by comparing it with mortality due to warfare. There were about 100 million fatalities in wars between the years 1700 and 1987. That yields a long-term average of about 350,000 fatalities per year. The yearly average between 1986 and 1991 has been estimated at about 427,800. These figures can be compared to the more than 12 million children's deaths in each of these years (Kent 1995).

The most lethal war in all of human history was World War II, during which there were about 15 million battle deaths. If civilian deaths are added in, including genocide and other forms of mass murder, the number of deaths in and around World War II was around 51,358,000. Annualized

Table 11.1 Annual Deaths of Children Under Five, Selected Years

Year	Child Deaths
1960	18,900,000
1970	17,400,000
1980	14,700,000
1990	12,700,000
1991	12,821,000
1992	13,191,000
1993	13,272,000
1994	12,588,000
1995	12,465,000
1996	11,694,000

Source: UNICEF, *The State of the World's Children* (New York: Oxford University Press/UNICEF, various years).

for the six-year period, the rate comes to about 8.6 million deaths a year—when children's deaths were running at well over 25 million per year. This most intense war in history resulted in a lower death rate, over a very limited period, than results from children's mortality year in and year out.

Counting late additions, at the end of 1987 there were 58,156 names on the Vietnam Veterans' Memorial in Washington, D.C. That is less than the number of children under the age of five who die every two days throughout the world. A memorial for those children who die worldwide each year would be more than 250 times as long as the Vietnam Veterans' Memorial, and a new one would be needed every year.

Children die for many different reasons. The immediate cause of death for most children is not murder, direct physical abuse, or incurable diseases such as AIDS but, as shown in Table 11.2, a combination of malnutrition and quite ordinary, manageable diseases such as diarrhea, malaria, and measles.

■ CHILDREN'S RIGHTS

Many different kinds of service programs are offered by both governmental and nongovernmental agencies to address children's concerns, and many of them have been very effective. However, the coverage is uneven, largely a matter of charity and chance. There is now an evolving understanding that if children everywhere are to be treated well, it must be recognized that they have specific rights to good treatment. Thus, there is now a vigorous movement to recognize and ensure the realization of children's rights.

Children's rights have been addressed in many different international instruments. On February 23, 1923, the General Council of the Union for

Table 11.2   Estimated Annual Deaths of Children Under Five, by Cause, 1986

Cause	Number (millions)	Proportion (%)
Diarrhea	5.0	35.4
Malaria	3.0	21.3
Measles	2.1	14.9
Neonatal tetanus	0.8	5.7
Pertussis (whooping cough)	0.6	4.3
Other acute respiratory infections	1.3	9.2
Other	1.3	9.2
Estimated total	14.1	100.0

Source: UNICEF, *The State of the World's Children 1987* (New York: UNICEF/Oxford University Press, 1987), p. 111.

Child Welfare adopted the Declaration of Geneva on the rights of the child. On September 26, 1924, it was adopted by the League of Nations as the Geneva Declaration on the Rights of the Child. It was then revised and became the basis of the Declaration of the Rights of the Child adopted without dissent by the UN General Assembly in 1959. The declaration enumerates ten principles regarding the rights of the child. As a nonbinding declaration, it does not provide any basis for implementation of those principles.

The Universal Declaration of Human Rights was approved unanimously by the UN General Assembly in 1948. It was given effect in the International Covenant on Civil and Political Rights and the International Covenant on Economic, Social, and Cultural Rights. The two covenants were adopted in 1966 and entered into force in 1976. The covenants include specific references to children's rights.

After ten years of hard negotiations in a working group of the UN Commission on Human Rights, on November 20, 1989, the UN General Assembly by consensus adopted the new Convention on the Rights of the Child. It came into force on September 2, 1990, when it was ratified by the twentieth nation. Weaving together the scattered threads of earlier international statements of the rights of children, the convention's articles cover civil, political, economic, social, and cultural rights. It includes not only basic survival requirements such as food, clean water, and health care, but also rights of protection against abuse, neglect, and exploitation, and the right to education and to participation in social, religious, political, and economic activities.

The convention is a comprehensive legal instrument, legally binding on all nations that accept it. The articles specify what states are obligated to do under different conditions. National governments that agree to be bound by the convention have the major responsibility for its implementation. To provide added international pressure for responsible implementation, Article 43 calls for the creation of a Committee on the Rights of the Child. It consists of ten experts whose main functions are to receive and transmit reports on the status of children's rights. Article 44 requires states parties to submit "reports on the measures they have adopted which give effect to the rights recognized herein and on the progress made on the enjoyment of those rights." Article 46 entitles UNICEF and other agencies to work with the committee within the scope of their mandates.

By the middle of 1997, all countries except Somalia and the United States had ratified or otherwise acceded to the Convention on the Rights of the Child. Somalia has not ratified because it does not have a functional government. The reasons for the United States' failure to ratify is not so clear. Both Bill and Hillary Clinton were known as strong child advocates when Clinton first took office in 1993, so it was a serious disappointment to children's advocates when the Convention on the Rights of the Child was not quickly signed and ratified. The United States finally did sign the convention in February 1995. That signing, handled very quietly, apparently was done to fulfill a deathbed promise to James Grant, who had been executive director of UNICEF. However, the convention does not become binding on the United States until it is ratified through the advice and consent of the Senate. As of mid-1998 the convention still had not been forwarded to the Senate for its action.

The U.S. government has never offered any official explanation for its reluctance to ratify the Convention on the Rights of the Child. However, the major objections that have been voiced unofficially have been these:

1. *States rights.* The historical struggle to find an appropriate balance between the powers of the states and the power of the national government has not been fully resolved. There is a fear that through its power to make international agreements, the U.S. government might federalize issues that previously had been addressed only in state law.

2. *Capital punishment.* Article 37 of the convention states that "neither capital punishment nor life imprisonment without possibility of release shall be imposed for offenses committed by persons below eighteen years of age." Along with Pakistan, Iran, Iraq, and Bangladesh, the United States is one of the few remaining countries that execute people for crimes committed before their eighteenth birthday. This is tied in with the argument that capital punishment should be a matter of state rather than federal policy.

3. *Abortion.* The preamble of the convention says that "as indicated in the Declaration of the Rights of the Child adopted by the General Assembly of the United Nations on November 20, 1989, the child, by reason of his physical and mental immaturity, needs special safeguards and care, including appropriate legal protection before as well as after birth." The last six words conform to the prolife, antiabortion position. However, because of the divisiveness of the abortion issue, the drafters of the convention chose not to elaborate the theme. For prolife activists, the convention is not explicit enough regarding safeguards before birth and thus is not acceptable.

It would be possible for the United States to ratify the convention despite these objections. It could be ratified with a reservation regarding the capital punishment provision, thus reserving the United States' freedom on that issue. There is nothing in the convention that would constrain the United States on the abortion question. Thus, there is no reason the United States should forgo supporting all the other provisions of the convention that it favors because of these objections.

Perhaps the most serious obstacle to U.S. ratification is ideological. The United States tends to support civil and political rights but not economic and social rights. In the United States there is no constitutional right to food or housing, for example. The Convention on the Rights of the Child includes social and economic rights of the sort that trouble many people in the government.

Conservative elements in the United States have organized systematic campaigns of opposition to the convention based on false charges that the convention would undermine the family and take away parents' right to raise their own children as they see fit. These unfounded arguments are advanced by people who apparently have not read the convention.

■ INTERNATIONAL OBLIGATIONS

Children have not only the rights enumerated in the Convention on the Rights of the Child but also, with few exceptions, all other human rights. The articulation of these rights in international instruments represents an important advance, but there is still much more to be done to ensure that these rights are fully realized. Although technically binding on the states that ratify these international human rights agreements, the human rights claims are ambiguous. The ambiguity in the international instruments is to some extent deliberate, because it is left to the national governments, representing the states that are parties to the agreements, to concretize them in ways appropriate to their particular local circumstances.

Much more needs to be done by national governments to ensure that human rights within their jurisdictions are realized. However, the question remains: What are the obligations of the international community, especially where national governments are unwilling or unable to do what needs to be done to ensure that human rights are fully realized?

There are programs of international humanitarian assistance and many international organizations, governmental and nongovernmental, that work to alleviate suffering. Development and foreign aid programs do a good deal to improve the quality of life. But it is now largely a matter of politics and charity. There may be a sense of moral responsibility, but there is no sense of legal obligation, no sense that those who receive assistance are entitled to it and that those who provide it owe it. Historically, the idea of a duty to provide social services and to look after the weakest elements in society has been understood as something undertaken at the national and local levels, not as something that ought to be undertaken globally. Indeed, the only major market economy in which there is no clearly acknowledged responsibility of the rich with respect to the poor is the global economy.

Within nations, citizens may grumble when they are taxed to pay for food stamps for their poor, but they pay. Globally, there is nothing like a regular tax obligation through which the rich provide sustenance to the poor in other nations. The humanitarian instinct and sense of responsibility is extending worldwide, but there is still little clarity as to where duties lie. There is no firm sense of sustained obligation at the global level.

Most current discussions of global governance focus on security issues, the major preoccupation of the powerful, and give too little attention to the need to ensure the well-being of ordinary people. Just as there should be clear legal obligations to assist the weak in society at the local and national levels, those sorts of obligations should be recognized at the global level as well. Discussion of that idea has begun in the United Nations, but just barely.

There is much discussion of international protection of human rights, but what does that mean? If one party has a right to something, some other party must have the duty to provide it. Children's rights would really be international only if, upon failure of a national government to do what was necessary to fulfill those rights, the international community was *obligated* to step in to do what needed to be done—with no excuses. There is now no mechanism and no commitment to do that. The international community provides humanitarian assistance in many different circumstances, but it is not required to do so. International law does not now require any nation to respond to requests for assistance.

There should be clear global obligations, codified in explicit law, to sustain and protect those who are the worst off. The exact nature of those obligations and their magnitude and form will have to be debated, but the debate must begin with the question of principle. The principle advocated here is that *international humanitarian assistance should be regularized through the systematic articulation of international rights and obligations regarding assistance.* Regularization can begin with the formulation of guidelines and basic principles and then perhaps of agreed codes of conduct. These can be viewed as possible precursors of law.

The nations of the world could collectively agree that certain kinds of international assistance programs *must* be provided, say, to children in nations in which children's mortality rates exceed a certain level. This international obligation to provide assistance should stand unconditionally where national governments or, more generally, those in power consent to receiving the assistance. The obligation must be mitigated, however, where those in power refuse the assistance and delivering the assistance would require facing extraordinary risks.

Part of the effort could focus on helping nations ensure that their children's nutrition rights are realized. The most prominent international governmental organizations (IGOs) concerned with nutrition are the Food and Agriculture Organization (FAO) of the United Nations, the World Food

Programme (WFP), the International Fund for Agricultural Development (IFAD), the World Health Organization (WHO), and UNICEF. They are governed by boards composed of member states. Responsibility for coordinating nutrition activities among these and other IGOs in the UN system rests with the Administrative Committee on Coordination/Subcommittee on Nutrition (ACC/SCN). Representatives of bilateral donor agencies such as the Swedish International Development Agency (SIDA) and the United States Agency for International Development (USAID) also participate in ACC/SCN activities. There are also numerous international nongovernmental organizations (INGOs) concerned with nutrition.

The main role of the IGOs is not to feed people directly but to help nations use their own resources more effectively. A new regime of international nutrition rights would not involve massive international transfers of food. Its main function would be to press and help national governments address the problem of malnutrition among their own people, using the food, care, and health resources within their own nations. There may always be a need for a global emergency food facility to help in emergency situations that are beyond the capacity of individual nations, but a different kind of design is needed for dealing with chronic malnutrition. Moreover, as chronic malnutrition is addressed more effectively, nations would increase their capacity for dealing with emergency situations on their own. Over time the need for emergency assistance from the outside would decline.

The IGOs could be especially generous in providing assistance to those nations that create national laws and national agencies devoted to implementing nutrition rights. Poor nations that are relieved of some of the burden of providing material resources would be more willing to create programs for recognizing nutrition rights. Such pledges by international agencies could be viewed as a precursor to recognition of a genuine international duty to recognize and effectively implement rights to adequate nutrition.

Of course, the objective of ending children's malnutrition in the world by establishing a regime of hard international nutrition rights is idealistic. Nevertheless, the idea can be useful in setting the direction of action. We can think of the IGOs as having specific duties with regard to the fulfillment of nutrition rights. We can move progressively toward the ideal by inviting IGOs to establish clear rules and procedures they would follow *as if* they were firm duties.

■ CONCLUSION

Within nations, through democratic processes managed by the state, some moral responsibilities become legal obligations. A similar process is needed at the global level. Internationally recognized and implemented rights and obligations should not and, realistically, cannot be imposed.

They should be established democratically, through agreement of the nations of the world. Reaching such agreement would be action not against sovereignty but against global anarchy. It is important to move toward a global rule of law.

Regularized assistance to the needy under the law is a mark of civilization *within* nations. If we are to civilize relations *among* nations, international humanitarian assistance also should be governed by the rule of law. Looking after our children internationally could become the leading edge of the project of civilizing the world order.

■ QUESTIONS

1. Should countries be concerned with the treatment of children within other countries? Should the United Nations be concerned? Explain.

2. Should corporations be allowed to benefit from exploitative child labor in other countries? Explain.

3. Should the United States ratify the Convention on the Rights of the Child? Why or why not?

4. Do you agree with the concept of *international obligations* for children's rights?

■ SUGGESTED READINGS

Freeman, Michael, and Philip Veerman, eds. (1992) *The Ideologies of Children's Rights*. Drodrecht, Netherlands: Martinus Nijhoff.
International Journal of Children's Rights (quarterly).
Kent, George (1991) *The Politics of Children's Survival*. New York: Praeger.
——— (1995) *Children in the International Political Economy*. London: Macmillan; New York: St. Martin's Press.
LeBlanc, Lawrence J. (1995) *The Convention on the Rights of the Child: United Nations Lawmaking on Human Rights*. Lincoln: University of Nebraska.
Sawyer, Roger (1988) *Children Enslaved*. London: Routledge.
United Nations (1992) *Child Mortality Since the 1960s: A Database for Developing Countries*. New York: United Nations.
United Nations Children's Fund (annual) *The Progress of Nations*. New York: UNICEF.
United Nations Children's Fund (annual) *The State of the World's Children*. New York: Oxford University Press/UNICEF.
United Nations Children's Fund (1993) *Food, Health and Care: The UNICEF Vision and Strategy for a World Free from Hunger and Malnutrition*. New York: UNICEF.
Veerman, Philip E. (1992) *The Rights of the Child and the Changing Image of Childhood*. Dordrecht, Netherlands: Martinus Nijhoff.

Health

Marjorie E. Nelson

Today our global community is more urban than village. Almost half the world's people live in urban areas, and more than 10 percent live in cities with a population of more than 1 million (WCED 1987). Forty years ago, our metaphor for human speed was Superman flying "faster than a speeding bullet." Now our metaphors are "warp speed" and "transporters." We can fly from Miami to Managua in three hours, from Palo Alto to Beijing in thirteen, and from New York to Nairobi in fifteen. Any two persons in our global community are separated by fewer than twenty-four hours travel. What happens in one "neighborhood" of our global community can affect us rapidly, even though we live in another.

■ NUTRITION AND HEALTH

The World Health Organization (WHO) defines health as a "state of complete physical, mental, and social well-being and not merely the absence of disease or infirmity." It is the result of a person's interactions with the environment. Genes, resources, past experiences, and choices determine some of the interactions humans have with their environment. The makeup of the natural environment is also a determinant. The complex interplay of all these factors results in a unique experience of health and disease throughout an individual's life. We examine some of these factors in this chapter. First, we look at nutritional status; second, we review examples of environmental factors; third, we examine some human behavioral factors; and finally, we consider some specific societal responses to protect health.

"Phuong caught measles last week. He's much thinner than usual. His mother says he has no appetite and she thinks he's going to die," Madame Xuan Lan told me.

It was 1968. I was making my weekly visit to the day care center. Phuong, his mother, and three sisters were recent arrivals in one of the refugee camps around Quang Ngai City, where the American Friends Service Committee had set up this center during the Vietnam War. Phuong's mother supported her family by gathering wild greens to sell each day. She left Phuong in the care of his eight-year-old sister while she went to market.

While Xuan Lan went for Phuong, I examined other sick children. When Phuong arrived he was too weak to stand. I could hear pneumonia in his lungs and he already had a pressure ulcer on his lower back. His mother was right: lying without food or medicine on a damp floor in a refugee camp, Phuong would soon be dead.

We bandaged the ulcer and I left medicines for the pneumonia. "Bring him to school every day. Feed him chao thit ga three times a day and just let him rest." Chao thit ga is a Vietnamese chicken rice soup. It would give him the calories, protein, and fluids he needed to fight the infection, heal his ulcer, and regain weight. Sure enough, soon his pneumonia and ulcer were gone, and he was the bright-eyed four-year-old we had known before measles struck.

In North America, we do not think of measles as a killer. However, in developing countries, about one out of every four children who gets measles will die. The main reason is malnutrition.

◼ Undernutrition

Adequate nutrition is crucial for a child's health. A good diet must include energy (calories), protein, and key micronutrients. Without adequate calories, a child's growth is stunted and weight is reduced. In case of infection, or in a time of food scarcity, a child like Phuong has no reserves on which to draw. The body needs protein for cell growth and repair. While calories are the fuel, proteins are the building blocks of the body; enzymes, hormones, cell walls, and antibodies fight infections. Micronutrients such as iron, iodine, and vitamins are also important.

In 1995, according to the United Nations Children's Fund (UNICEF), 165 million children under age five were malnourished—half of them in just three countries: Bangladesh, India, and Pakistan (1996b: 12). Malnourished children, if they survive, grow into stunted adults with increased risk of disease, lowered capacity for physical and mental work, and, for women, reduced ability to nourish a developing fetus during pregnancy.

Many of these effects cannot be totally reversed, even if nutrition improves later.

■ Interaction of People, Land, and Food

Nutrition depends on what we eat. The interaction of people with the land at their disposal affects food availability and hence nutrition. When a people outgrows its food supply, one of three things will happen: agricultural change will be introduced to produce more calories and protein per acre of arable land (land that can be cultivated); more land will be acquired—by migration or conquest; or people will starve.

In a traditional diet (a diet eaten by a people over a period of 500–1,000 years), the main source of calories is from grains and starchy foods: for example, wheat, potatoes, rice, beans, corn, plantains, and cassava. The highest-quality protein comes from animal sources: meat, milk, eggs, and seafood. However, other substantial protein sources for much of the world are soybeans, peas, beans, wheat, corn, rice, oats, rye, nuts, and especially peanuts. In traditional diets, fat is valued for its calories and taste but is a minor part of the diet.

To provide the yearly protein supply for a moderately active man by beef alone would require about 4.75 acres of land. To meet his yearly needs with poultry alone would require just under 2 acres. About 0.7 acre of wheat, 0.5 acre of rice, or 0.2 acre of soybeans would also meet his requirements. As the amount of arable land per person decreases, the usual diet in that region will shift toward food that has a higher protein content per acre.

Japan, a densely populated island nation, has a traditional diet that depends heavily on rice, soybeans, and seafood—which, coming from the sea, does not depend on arable land. Why? At the end of World War II, Japan had about 0.15 acre of arable land per person; this was not enough to meet their people's protein needs, even if it were *all* in soybeans. Today it is about 0.12 acre per person. Japan, one of the richest and most developed countries, will always have to import food as long as its population remains as large as it is now. But the traditional Japanese diet tells us the country has had this problem for much longer than fifty years.

Rice was grown in Japan but was not significant in the Japanese diet until the seventeenth century. From 1600 to 1868, Japan underwent great economic development, and its population rose to 30 million. "Progress on this scale was possible only because of a constant rise in the agricultural production which supported these 30 million people on a surface area which could only have supported five or ten million people in Europe." Rice, a concentrated source of calories and protein, was the key (Braudel 1981: 156).

Now consider an example of migration/starvation. Food, as well as people, can migrate. The potato is a Western Hemisphere food. Grown in the high Andes since at least 2000 B.C.E., Spaniards brought it to Europe in the mid-1500s where it joined wheat (bread) as a major food. Soon it was widely grown in Ireland, and by the 1700s it was almost the entire diet of Irish peasants, augmented only by a little milk and cheese. By 1843, it was the single food sustaining the population of Ireland. Then disaster struck. A new potato blight appeared and wiped out the potato crop for two successive years. For many reasons, adequate food was not brought into Ireland. Thousands died of starvation, while Irish immigration to North America increased greatly.

The typical U.S. diet is often described as "bread, meat, and potatoes." This could be a pot roast, bread, and baked potatoes; a hamburger patty, bun, and fries; or fried chicken with mashed potatoes, gravy, and roll. Traditionally, the U.S. consumer has expected to dine on meat with potatoes and bread on the side. At the end of World War II, the United States had 2.9 acres of arable land per person. Today it is about 2.1 acres per person. We are a nation of immigrants who have come over a period of 350 years to a place with abundant arable land. Our diet reflects that, as well as the diet patterns of our countries of origin. This brings us to another kind of malnutrition: overnutrition.

■ Overnutrition

In the face of limited or sporadic food supply, homo sapiens developed the ability to store calories as fat when there was a temporary excess of food. Dietary fat and carbohydrates can be converted to body fat and stored. Gram for gram, fat contains twice as many calories as carbohydrates, so it is an efficient energy storage system. In our distant past, persons with a preference for calorie-rich foods and an efficient system for storing extra calories as fat were more likely to survive, nourish offspring, and pass on both dietary preferences and genetic traits. Today in some areas of the world, food surpluses, high incomes, and technology for the mass processing and distribution of food make excess food available to many people. These foods are often high in fat and simple sugars and low in fiber compared to traditional unprocessed foods such as fresh fruits, vegetables, and grains. As a result, obesity is a growing problem in most developed countries. Obesity is 120 percent or more of ideal body weight (see Table 12.1). Today nearly 33 percent of the U.S. population is obese.

Weight gain usually begins in childhood or early adulthood and continues gradually throughout life. In recent years, lack of exercise has added to the risks from an unhealthy diet. Obesity and the diet that leads to it cause serious health problems: high cholesterol, high blood pressure, and

Table 12.1   Desirable Weight, in Pounds, for Adults Age Twenty-five and Over
(indoor clothing)

Height (in shoes)	Small Frame	Medium Frame	Large Frame
Men			
5 ft. 2 in.	112–120	118–129	126–141
5 3	115–123	121–133	129–144
5 4	118–126	124–136	132–148
5 5	121–129	127–139	135–152
5 6	124–133	130–143	138–156
5 7	128–137	134–147	142–161
5 8	132–141	138–152	147–166
5 9	136–145	142–156	151–170
5 10	140–150	146–160	155–174
5 11	144–154	150–165	159–179
6 0	148–158	154–170	164–184
6 1	152–162	158–175	168–189
6 2	156–167	162–180	173–194
6 3	160–171	167–185	178–199
6 4	164–175	172–190	182–204
Women			
4 ft. 10 in.	92–98	96–107	104–119
4 11	94–101	98–110	106–122
5 0	96–104	101–113	109–125
5 1	99–107	104–116	112–128
5 2	102–110	107–119	115–131
5 3	105–113	110–122	118–134
5 4	108–116	113–126	121–138
5 5	111–119	116–130	125–142
5 6	114–123	120–135	129–146
5 7	118–127	124–139	133–150
5 8	122–131	128–143	137–154
5 9	126–135	132–147	141–158
5 10	130–140	136–151	145–163
5 11	134–144	140–155	149–168
6 0	138–148	144–159	153–173

Source: Reprinted courtesy of Metropolitan Insruance Company, *Statistical Bulletin.*

hardening of the arteries. This results in heart attacks and strokes, the second and third leading causes of death in the United States. Gallstones and diabetes are more common in obese people. High-fat, high-calorie diets are also often low in dietary fiber. Dietary fiber, found in whole grains, fruits, and vegetables, slows the absorption of fat and sugar from the digestive system. This helps counteract a high-fat, high-calorie intake. Fiber affects bacterial metabolism in the digestive system, thus reducing the absorption of toxins. It also has a laxative effect. All these actions of fiber reduce the risk of some cancers, especially colon cancer. Cancer is the leading cause of death in the United States, and colon cancer is one of the top three along with breast and lung cancer. Once obesity occurs, like undernutrition, it is very hard to reverse the effects. The best strategy is to form good diet and exercise habits early and maintain them throughout life.

■ NATURAL ENVIRONMENT: NEW AND REEMERGING DISEASES

In 1959, two sailors died of a rare disease: pneumocystis pneumonia. One was in England; the other, a Haitian, was in New York. About that time, Dr. Margrethe Rask left Denmark to work as a surgeon in Zaire, and Gaetan, a boy in Quebec, dreamed of a career as an airline steward. In 1966, a Norwegian sailor died of multiple infections and a strange immune system collapse. Later, his wife and one child also died from results of severe immune deficiency. Two years later Robert, a teenager in St. Louis, died of multiple infections and Kaposi sarcoma. He had never been abroad, but he had been heterosexually active for "several years."

During the 1970s, Dr. Rask suffered from general lymph node swelling and increasing fatigue. Gaetan, who spoke only French, learned English and got a job with Air Canada. He became sexually active as a homosexual. He flew often between France, the United States, and Canada. A young Greek fisherman moved to Zaire where he worked on Lake Tanganyika, while each year several thousand men from Haiti went to Zaire as short-term laborers before returning home. On July 4, 1976, New York harbor was filled with the tall sails of ships from fifty-five nations and their sailors who had gathered to help the United States celebrate its 200th birthday. It was a glorious party. In Zaire later that year, Dr. Rask collapsed with her strange fatigue and decided to return to Denmark for treatment. The Greek fisherman was not feeling as healthy as usual either. In late 1977, Dr. Rask died at the age of forty-seven with overwhelming pneumocystis pneumonia. Meanwhile, Kiyoshi Takatsuki discovered people on some islands of Japan who had a cancer of the T-cells of their immune systems. Both Japanese and U.S. research laboratories isolated a retrovirus from their cells. It was also identified in monkeys from Japan, Indonesia, and Kenya. In 1978, Dr. Peter Piot, a tropical disease specialist in Belgium, saw a new patient, a Greek fisherman in his late thirties from Zaire. The man died of widespread infection with an odd mycobacterium— one not recognized as a disease agent.

Meanwhile, Gaetan flew to San Francisco for the Gay Freedom Day parade. He enjoyed the visit and decided to return often. The next year, Gaetan experienced a "flu-like" illness that left him with swollen lymph nodes all over. He continued to fly between Paris, New York, and San Francisco, where he had many sexual contacts. In 1980, his doctor told him he had a rare cancer: Kaposi sarcoma. In Belgium, Dr. Piot continued to see patients from Zaire with severely damaged immune systems and strange infections: some women, some men, and some married couples.

On April 28, 1981, Sandra Ford, a technician at the Centers for Disease Control (CDC) in Atlanta, was puzzled. Part of her job was to fill orders for pentamidine, a drug used so rarely it was stocked only at the CDC. It was used to treat a rare pneumonia, pneumocystis, which typically occurred when a disease knocked out a patient's immune system, such as a child with leukemia or a patient with an organ transplant. Usually she got a dozen or so requests per year, but already this year she had nine requests, and the doctors did not seem to know why their patients had pneumocystis. Several were in New York City. She wrote her boss a memo about this strange pattern. Three weeks later, a doctor from Los Angeles called Dr. Mary Guinan at the CDC because he and his colleagues had a group of cases of pneumocystis pneumonia in gay men. They asked to have a report of this cluster appear in the CDC's weekly bulletin, MMWR. *The June 5, 1981, issue of* MMWR *carried a report titled "Pneumocystis pneumonia—Los Angeles." This was the first published report of a group of cases in the worldwide epidemic we have come to know as AIDS.* (Based on Shilts 1987 and Garrett 1994)

Technically an epidemic is "the unusual occurrence of a disease in the light of past experience." Sandra Ford knew that even nine cases of unexplained pneumocystis pneumonia was a possible epidemic. When epidemiologists investigate a possible epidemic, they look for "Patient Zero." If they can find the first patient in the epidemic and learn about his or her interactions with other cases, they may find out how it spreads. If it is a new disease, it may help explain what causes it.

When CDC epidemiologists first began investigating what we now know as human immunodeficiency virus (HIV), the underlying cause of acquired immunodeficiency syndrome (AIDS), they did not know it was a virus. They also did not know how people became infected. They did know that it destroys part of the human immune system, allowing normally harmless microbes to invade and kill people. When they first found evidence that it was an infectious agent and was passed through sexual contact, they did not know how long a person could be infected before symptoms would occur. The first estimate was months. We now know a person may carry the virus and be infectious for more than ten to fifteen years before symptoms of AIDS appear.

In the early investigations, the CDC team found that Gaetan had been a sexual partner of several of the earliest cases in both New York and California. As they still thought the asymptomatic infectious period was months, he was tentatively labeled Patient Zero. In a 1982 interview, Gaetan estimated he had had more than 2,000 sexual contacts in cities of

Europe and North America. He continued having unprotected sex even after being warned by several doctors that he was probably infectious. He certainly contributed to the spread of the AIDS epidemic, but he was not Patient Zero. Many of the earliest cases of AIDS detected in gay men in both New York and California were in a small group of men who lived and partied together in New York during the 1976 bicentennial celebration. Perhaps HIV was introduced to that group then, three years before Gaetan presumably got his infection.

Slowly we have put together the pieces of this AIDS puzzle. When HIV tests became available, doctors went back and tested samples of most of the early unidentified cases. They were HIV positive. The retrovirus from Japan was a close relative, but it was not HIV. We may never know where and how the AIDS epidemic started for sure, but clearly some cases occurred as early as 1959. However, the epidemic took off in the 1970s. Look at the cases before 1975: three sailors, the wife and child of one of them, a missionary doctor, a heterosexually active teenager who never left home, and a Greek fisherman working abroad. This is not the pattern we associate with AIDS today. But if someone had had all those cases collected then, that person could have warned us that an epidemic was coming.

Because air travel was uncommon before the 1970s, most of the early cases were travelers by sea; and their trips lasted days or weeks. In 1950, there were 2 million international commercial air passengers. This grew to 74 million in 1970, 280 million in 1990, and will approach 600 million by the year 2000 (IATA annual report). People in our global community are now more mobile, and they move between large cities more quickly. In addition, there are more people in those cities than there were in 1950. Also, as the world population grows, we are expanding into parts of the planet where few people used to go.

Humans are not the only ones interacting with the environment in our global neighborhoods; each neighborhood is home to many other living organisms: animals, plants, and microbes. Sometimes, as in the case of the HIV virus, what one microbe does improves the climate for other microbes. When HIV viruses have destroyed enough T-cells in a person's body, microbes, which are usually killed by our T-cells, can grow. Mycobacteria such as tuberculosis, parasites such as pneumocystis, or a new herpes virus that may be linked to Kaposi sarcoma are all examples.

Viruses and bacteria not only mutate—that is, change their genetic structure—but also exchange "tools" with other microbes. Many of these changes are useless or harmful to the microbe, but sometimes they help. One tool a bacteria might have is resistance to an antibiotic. Some bacteria carry the information on how to do this in "tool boxes" called plasmids, which they exchange with other bacteria. Before we discovered and started using antibiotics, this set of tools was not so useful, but in the past fifty years, bacteria that have one survive better.

Consider interaction of cholera bacteria with their environment. When we ingest water or food contaminated by cholera, they attach to the intestinal lining and reproduce. Unfortunately for us, they release a protein that causes severe diarrhea that can lead to death in days, or even hours.

Cholera has been present in South Asia for centuries. Periodically it would erupt as an epidemic across many countries, often carried by pilgrims traveling to Mecca. In the 1800s, there were many outbreaks in European cities due to inadequate sewage systems and unprotected water supplies. This virulent organism killed 70–80 percent of the people it infected, but it could not survive long outside the human body. Dr. John Snow, an early epidemiologist, stopped one outbreak in London by removing the handle from the pump at the Broad Street well—a cholera-contaminated water supply. Slowly, public health measures such as safe water, good sewage disposal, and use of quarantine reduced cholera epidemics.

Then, in 1961, a new strain of cholera, El Tor, appeared in Indonesia. It produced a milder disease; fewer ill people died, and some people were asymptomatic carriers. However, this strain was resistant to antibiotics. One substrain in Thailand was resistant to eight drugs. This cholera bacteria had picked up a powerful set of tools in a toolbox trade somewhere. It had also developed a way to survive longer in saltwater between stays in human hosts. These cholera organisms can hibernate for weeks at a time inside plankton that grow with algae. They can drift inside their hosts on the ocean currents or be drawn into a ship's bilgewater and be pumped out in the next port of call a continent away. Both international shipping and increased algal blooms in the past three decades have helped the spread of cholera. A fiercer strain, Bengal, which appeared in the 1990s, combined some of the advantages of the El Tor with the more virulent disease-causing strength of the classic cholera.

During the past fifty years, we have lived through the Golden Age of Antibiotics. Now we are seeing new diseases like AIDS and reemergent old diseases with new defenses, like Bengal cholera, against which our antibiotics are no longer effective. In the next century, we must find new ways to deal with them and their toolboxes of tricks. This will require worldwide cooperation, as well as respect for changes that may arise in the environment as a result of our actions or other interactions.

■ HEALTH AND HUMAN BEHAVIOR

Above I pointed out that past experiences and personal choices determine some of the interactions humans have with their environment and that the social characteristics of that environment determine some of those interactions. Both affect health.

Rapid air transportation was not the only human behavior factor that helped the spread of the HIV/AIDS epidemic. The development of reliable birth control and of antibiotics that cured many of the recognized sexually transmitted diseases contributed to a sexual revolution among both hetero-sexuals and homosexuals. Most AIDS cases in the world today result from heterosexual sexual activity.

In contrast, the first cluster of AIDS cases identified in the United States was among gay men. It occurred during the tenure of a politically conservative administration in Washington that was not sympathetic to this group of citizens, nor to public health experts who tried to get more re-sources allocated to the emerging epidemic. The history of AIDS control efforts might have been very different if the first cluster had been detected in recipients of blood transfusions, for example.

Both individual actions and social structures or attitudes affect pat-terns of disease. These are not limited by national borders. Decisions in an Asian or Colombian neighborhood of our global community affect people in distant countries. During the Vietnam War, many soldiers were intro-duced to cheap heroin in Southeast Asia. The flow of heroin from there on both military and commercial flights rose dramatically. The 1970s saw the beginning of our current epidemic of intravenous (IV) drug use. In Latin America, a more powerful formulation of cocaine, "crack," was invented and began to spread among drug users. IV drug and crack cocaine users also have high rates of AIDS.

Not all drug traffic flows from developing to developed countries. Some is home grown and some flows the other way. In fact, the United States might be said to be the largest "drug pusher" country in the world if we consider the case of tobacco.

Lucy was a coworker and a patient of mine. We had known each other for years. Now she sat in my office with bad bronchitis.

"Lucy, have you ever thought about giving up the cigarettes?" I asked. She had smoked two packs a day since she was a teenager.

She swung one foot to and fro. "I've tried lots of times."

"Stop-smoking classes or nicotine gum have helped a lot of people quit," I suggested.

Her shoulders drooped. "I know, but I just can't quit. I guess I'll just have to die early."

I gave her a prescription for her bronchitis. I felt sad as I watched her walk down the hall. Her last words seemed ominous.

A couple of years later, Lucy had a heart attack and fell, breaking her hip. It was not clear which happened first, but she died suddenly in the hospital from complications of the two events. She was fifty-two. Her words in my office that day had been prophetic: dying at fifty-two is quite early for a North American woman.

We are in the middle of an epidemic of disease and death caused by tobacco. It is a slow-growing epidemic compared to a cholera epidemic, which develops in days, or even to the AIDS epidemic. It takes thirty to thirty-five years for the damage caused by tobacco to show up as illness or death. Cigarette smoking causes lung cancer as well as heart attacks and thinning of the bones. Figure 12.1 shows that we really have two epidemics: one in men and one in women.

The epidemic began to show up in men in the 1940s but not in women until the late 1960s. Why? Smoking was not widespread in the United States before World War I (1914–1918). However, one of the standard items given to soldiers in World War I, along with equipment and food rations, was cigarettes. Every soldier got them. Many men came back with "the habit" and continued smoking. But it was a man's habit. It was not socially acceptable for women to smoke until World War II (1940–1945). So we see a twenty-five-year lag between the lung cancer epidemics in U.S. men and women.

This same pattern is emerging around the world, due largely to U.S. and European tobacco companies. While tobacco is widely grown, in traditional societies it was the older men who used it. They grew it locally, cured it themselves, and "rolled their own" one at a time. If a man did not start smoking until he was a mature adult, he might live out his expected life span before he developed a cancer or died of other complications of smoking.

Figure 12.1   Cancer Death Rates in U.S. Women and Men, 1930–1993

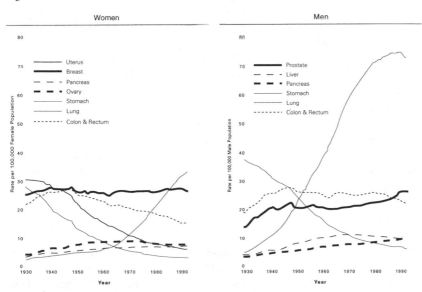

Source: Reprinted by the permission of the American Cancer Society, Inc.

Today tobacco is the world's most widely grown nonfood crop and a significant item in international trade. According to the International Tobacco Growers Association, eighteen countries account for 80 percent of tobacco produced for export. The United States heads that list, exporting 30 percent of its total production. Asian countries, due to their rapid economic growth, are prime targets for U.S. cigarettes. Both the Reagan and Bush administrations, for example, used the threat of trade sanctions to help open markets in Japan, Taiwan, and South Korea for U.S. tobacco companies.

In the United States, more money is spent on advertising tobacco products than on anything else except automobiles. In 1993, tobacco companies spent more than $6 billion on advertising and promotional items (Altman et al. 1996). The results? Nearly one in five U.S. high school seniors smokes daily, and about 40 percent of those smokers are female. Like Lucy, most adult smokers acquire their habit in their teens.

In China and Vietnam, where Western tobacco companies and their advertising are not yet well established, the traditional pattern predominates. Old men smoke and younger men are starting, but it is still uncommon among women, especially adolescent females. In Vietnam, only 4 percent of women smoke (Jenkins 1997), and in Beijing, only 3 percent of school-age girls smoke (Zhu 1996). In countries where Western tobacco companies have introduced advertising and promotion campaigns, smoking patterns become more like those of the United States and Europe.

The health effects of this increase in tobacco use are enormous. In addition to lung cancer, tobacco use also causes many deaths from emphysema of the lungs. However, the greatest number of deaths due to tobacco use are cardiovascular: heart attack and stroke. Many of these deaths are "premature"—that is, before the age of sixty-five. Like Lucy, these people die in their forties and fifties. Tobacco use accounts for 27 percent of all premature deaths in the United States (Amler and Dull 1987). Already, lung cancer is the most common cancer in China. By early next century, China will have 2 million smoking-related deaths each year; 900,000 of them will stem from lung cancer alone (Zhu 1996).

Tobacco use is a good example of how both individual choice and social environment affect disease in our global community. International trade, advertising, social values in a "neighborhood," government policies, and the like, as well as an individual's decision, influence the magnitude and pattern of disease.

* * *

So far we have looked at examples in three categories that influence health and disease in humans. The first category was *nutrition*. We saw how Phuong's malnutrition left him with no reserves to fight against a serious illness. The second category was *natural environment*. We saw how a new disease, AIDS, appeared and became a worldwide epidemic. New microbes, or reemerging old ones like Bengal cholera, with new tools to survive and spread better, present constant challenges to our health. The third category was *social environment*. We saw how the cultivation, marketing, and use of tobacco products have produced epidemics of lung cancer and increased the rate of heart attacks, strokes, and emphysema, which lead to illness and early death.

■ PROVIDING AND PROTECTING HEALTH: COMMUNITY ACTIONS

In this final section, we consider how our global village responds to health challenges. What do we as a community do to protect our health?

▨ Maternal and Child Health

The nurses at Van Dinh District hospital were excited. Tuyet, a young woman from a small hamlet, had come to the hospital to have her first baby. Instead, she had given birth to three girls. Triplets are even more rare among Vietnamese than in the West. Everyone in the hospital wanted to see these three miracle babies. Although healthy, they were small so they would be in the hospital for several weeks. Dr. Thuat, the medical director, was pleased, not only because Tuyet and her daughters were healthy, but also because he saw an opportunity. When Tuyet went home, everyone would want to visit with her and the babies. She would be admired and respected because of these healthy triplets. Other young women would listen and take advice from her. So he went over to the obstetrics wing and invited Tuyet to enroll in their community health worker course. She could stay here with her babies while she studied. To his delight, Tuyet agreed. (See Figure 12.2.)

Every community values its children, for they are the future. If children do not survive and grow into strong healthy adults, the community grows weak and risks extinction. So every community has developed activities to protect the health of children and their mothers.

As we have seen, nutrition is crucial to health. Adequate nutrition must begin before a baby is born. If a mother is poorly nourished, the developing fetus will not get all the nutrients it needs. The baby may be born underweight, or even with birth defects. If a pregnant woman smokes, she may have a premature infant. If she drinks heavily, the baby may be retarded. Women need to know these things before they become pregnant, and they need access to good nutrition during pregnancy. In many countries, community health workers, like Tuyet, are trained to share this information with women in their neighborhood.

Figure 12.2   Healthy Eight-Month-Old Triplets

Photo Credit: Marjorie E. Nelson

These healthy eight-month-old triplets prove to be an attraction that helps their mother,
a community health worker, deliver health education messages to her village
in Van Dinh District, Vietnam (1973).

They encourage women to breast-feed their infants. Breast milk is the
best food for infants in their first four to six months of life for four reasons.
First, it contains all the essential nutrients the growing baby needs except
iron. Second, it contains antibodies to fight infections. A baby's immune
system is not fully developed until about six months of age. Until then, it
must depend on antibodies in mother's milk. The third and fourth reasons
breast milk is best are that it is sterile and inexpensive. If a baby is fed with
formula or other foods, there is a higher chance of illness, because poor
families often stretch formula or other foods with contaminated water.

Sometimes community programs provide food supplements for pregnant and nursing women and young children. In the 1980s, the government
of Tamil Nadu in southern India began a program of weighing children
every month from six months to three years of age. Any child who was
malnourished received extra food for ninety days. At the same time, their
mothers received nutrition education and were also offered extra food.
When the program began, 45–50 percent of the children needed extra food.
Eight years later that had dropped to 24 percent. "[This program] worked

in Tamil Nadu because the community nutrition workers were well trained and highly motivated, and because mothers came to understand the importance of feeding for healthy growth" (World Bank 1993: 80). In the United States, this kind of community education work is often found in local health departments or community prenatal programs. At a county level, the Women, Infants, and Children (WIC) Program, funded by the U.S. Department of Agriculture, provides supplemental food to eligible women and children.

Some programs improve the nutrition and health of the whole population. Goiter and impaired thyroid function due to lack of iodine in the diet used to be quite common in the United States. It has disappeared, thanks to iodized salt. In the early 1900s, rickets caused by Vitamin D deficiency was very common in the northern United States and Europe. Now it is rarely seen because Vitamin D is added to milk.

▓ Vaccines and Immunization Programs

Vaccination, or immunization, causes the body to produce antibodies against a specific bacteria or virus. This gives a person the ability to fight off an infection when exposed. In spite of his malnutrition, if Phuong had been immunized against measles, he would not have become sick and almost died.

In 1965, I heard a team of infectious disease specialists at the CDC present a plan to eradicate smallpox in ten years. The roomful of doctors from all over the world were skeptical. Never in history had humanity wiped out an infectious disease. Surely it could not be done in just ten years. In fact, it took them eleven.

On October 26, 1977, a Somali cook, Ali Maow Maalin, became the last case of wild smallpox in the world. This took a worldwide team effort. The United States contributed $30 million to the smallpox eradication campaign. Each year it saves about $360 million because it no longer has to vaccinate against, watch for, or treat any cases of smallpox in the United States.

The global community, through WHO and UNICEF, is now waging another ambitious campaign: Expanded Program of Immunization (EPI). The goal is to immunize 90 percent of the world's children against six diseases by the year 2000. When this program was begun in 1980, only 25 percent of the world's children were immunized against these diseases. By 1995, this had increased to 80 percent, and forty-four countries have already reached the goal of 90 percent. As a result, polio and measles may be eradicated. Perhaps these diseases will soon join smallpox as just a memory (UNICEF 1996b).

◼ Influencing Human Behavior

There is no vaccine against AIDS. Combined treatment with two or more medicines prolongs the life of AIDS patients, but it is not yet clear that this can cure AIDS. These drugs may cost more than $3,000 per month per person. Most of the estimated 9 million people infected with HIV by 1990 cannot afford that, so they will die early and without any substantial treatment. This is a great tragedy. However, an even greater tragedy will occur if we as a global community let the epidemic continue to spread. We now know how to prevent transmission of the HIV virus. The World Bank noted in its 1993 World Development Report that

> a comprehensive AIDS prevention program could check the growth of the disease. . . . Crucial elements in these strategies are providing information on how to avoid infection, promoting condom use, treating other sexually transmitted diseases, and reducing blood-borne transmission. . . . Young people, both in and out of school, need comprehensive education on reproduction and reproductive health issues. . . . All potential behavioral choices, including abstinence and condom use, should be presented. (World Bank 1993: 100)

In 1993, the global community spent about $1.5 billion on AIDS prevention. Schools, religious groups, voluntary agencies, and governments have all been involved in prevention programs.

No medical technology exists to undo the damage caused by tobacco use. By 2020, tobacco will kill more people than any single disease, even AIDS (Murray and Lopez 1996). However, it is another epidemic we know how to stop. Prevention can help both smokers and those who do not smoke. Smokers who stop cut their heart attack risk to the same as nonsmokers in one year and their added lung cancer risk in fifteen years. Lung destruction of emphysema stops when smoking stops. Of course, for those who never start, all these risks are avoided. Stop-smoking programs are offered by many hospitals, clinics, and health departments. Voluntary agencies, such as the American Heart Association and the Lung Association, sponsor such programs as well as programs to prevent young people from starting to smoke or using smokeless tobacco.

Government policy also influences decisions to start or continue smoking. Actions that have helped reduce tobacco use include raising taxes on tobacco products, banning or limiting advertising, requiring warning labels on cigarette packages, prohibiting sales to minors, banning cigarette vending machines accessible to minors, and forbidding the sale of single cigarettes or distribution of free samples. Singapore has banned tobacco ads since 1971, and currently about eighteen countries have a total or partial ban on such ads. By the early 1990s, more than eighty countries required warning labels on cigarette packages. Tobacco taxes are fairly common. When Papua New Guinea raised the tobacco tax 10 percent, consumption dropped by 7 percent (World Bank 1993). These taxes can be used to support prevention programs.

▓ Poverty and Health

So far we have examined some examples of specific activities designed to improve directly the health of people. Another factor mentioned earlier was the *resources* a person possesses. Poor people have more disease and poorer health than rich people. Studies show that this is true all over the world. With more money, people can buy more food, live in a place with safer water and sewage disposal, and get more education and medical care.

A poor country will usually have higher disease rates, higher death rates, and lower life expectancy than richer countries. Table 12.2 shows that, in general, as a country's per capita gross national product (GNP) falls, life expectancy falls, and death rates among infants and pregnant women and rates of tuberculosis rise. In a poor country, both individuals and the community at large lack resources to reduce death rates and improve the health status of people. "In 1984 few counties achieved an average life expectancy at birth of 70 years or more until gross national product per head approached $5000 per year" (Wilkinson 1992: 168).

Once a country reaches a certain level of economic development (about $5,000 per capita GNP in 1984), the *gap* between its richest and poorest people is a more important influence on the health of the nation than per capita GNP. As the gap between the richest and poorest gets smaller, life expectancy at birth increases. If the gap gets larger, life expectancy drops. This is based on studies in twelve countries of Western Europe and in the United States, Canada, Australia, and Japan. One might say, "As the rich get richer, the country gets sicker." Consider Great Britain and Japan. "In 1970, income distribution and life expectancy were similar in these two countries. . . . Since then they have diverged. Japan has the most [equal] income distribution of any country on record [and] has the highest life expectancy in the world. In Britain, on the other hand, income distribution has widened . . . and deaths among men and women aged 15–44 has increased" (Wilkinson 1992: 168).

The United States has the widest gap between poorest and richest families of any developed country. Government policies in terms of tax relief and other transfers can alter this gap. For example, without government intervention, France and the United States would both have about 25 percent of children living in poverty. After transfers and taxes, France's child poverty rate goes to 6.5 percent, while the United States' rate falls to 21.5 percent. Just four other developed countries have child poverty rates above

Table 12.2 Per Capita GNP and Selected Health Indicators, Selected Countries

Country	Per Capita GNP (1994 U.S.$)	Life Expectancy at Birth	Infant Mortality Rate[a]	Maternal Mortality Rate[b]	Tuberculosis Incidence Rate[b] (1990)
Sub-Saharan Africa					
South Africa	3,010	63	54	230	250
Kenya	260	59	67	650	140
Zambia	350	49	106	940	345
Ethiopia	130	48	130	1,400	155
Middle East and North Africa					
Israel	14,410	76	9	7	12
Egypt	710	61	59	170	78
Yemen	280	52	109	1,400	96
Asia and Pacific					
Japan	34,630	79	5	18	42
China	530	69	38	95	166
India	310	60	90	570	220
Nepal	200	53	101	1,500	167
Americas					
Canada	19,570	77	7	6	8
United States	25,860	76	9	12	10
Venezuela	2,760	70	34	120	44
Haiti	220	55	94	1,000	333
Europe					
Norway	26,480	77	8	6	8
Portugal	9,370	74	11	15	57
Russian Federation	2,650	69	20	25	56

Sources: UNICEF, *Progress of Nations* (New York: UNICEF, 1996); World Bank, *World Development Report* (New York: Oxford University Press, 1993).
Notes: a. per 1,000 live births
b. per 100,000 live births and per 100,000 population respectively.

10 percent, while ten have rates below 5 percent. Those ten all have lower death rates in infants and pregnant women, and all but Denmark have the same or higher life expectancy as the United States (UNICEF 1996b).

Health is one of the most precious possessions of any person. Here we have considered some of the determinants of health: nutrition, poverty, the natural environment, the society in which we live, and the decisions we make individually or collectively. We have seen examples of local programs that address these factors: feeding programs, immunizations, and stop-smoking classes. Both government agencies (such as health departments) and private voluntary agencies may deliver these services. Other factors can be operating in another neighborhood of our global community: emergence of a new virus, increased travel, tobacco trade, cocaine smuggling, government policies, and so on. Communications and cooperation between national and international agencies, such as the CDC, WHO, and UNICEF, play an increasingly important role in protecting our health when such factors are involved. No one nation could eradicate smallpox or monitor AIDS. Yet, a successful response to these challenges benefits us all.

■ QUESTIONS AND SUGGESTED ACTIVITIES

1. How many examples of AIDS prevention programs can you identify in your own community?

2. If you saw the movie *Outbreak*, how does it demonstrate the influence of environment and human behavior on disease?

3. Should health care be a right of all people?

4. Should governments regulate the sale and purchase of tobacco products?

5. How do health problems differ in the North and South? Which problems are more serious? Which should receive the greatest attention?

6. Visit your local health department or voluntary agencies. Find out about their programs. How are they addressing any of the issues discussed in this chapter? Do they need volunteers?

7. Using Altavista or some other search engine, see what you can find on the World Wide Web about food and calories, the tobacco epidemic, or emerging infectious diseases.

■ SUGGESTED READINGS

Braudel, Fernand (1981) *The Structures of Everyday Life: Civilization and Capitalism—15th–18th Centuries*. New York: Harper & Row.
Crosby, Alfred (1972) *The Columbian Exchange: Biological and Cultural Consequences of 1492*. Westport, CT: Greenwood Press.
Garrett, Laurie (1994) *The Coming Plague*. New York: Farrar, Straus & Giroux.
Hobhouse, Henry (1987) *Seeds of Change: Five Plants That Transformed Mankind*. New York: Harper & Row.
McNeill, William H. (1987) *Plagues and Peoples*, New York: Anchor Books.
Shilts, Randy (1987) *And the Band Played On*. New York: St. Martin's Press.
Tannahill, Reay (1988) *Food in History*. New York: Crown Publishers.
World Commission on Environment and Development (1987) *Our Common Future*. Oxford: Oxford University Press.

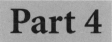

Part 4

The Environment

13

Protection of the Atmosphere

Mark Seis

In this chapter we explore global warming, ozone depletion, and acid rain. We examine each separately, looking at the cause of the problem, the major nation-state contributors, and the various international polices that have been proposed and implemented for reducing the potentially damaging effects each poses for the future of humanity.

■ THE THREAT OF GLOBAL WARMING

Since Charles Keeling set up laboratories in 1958 at the South Pole and Hawaii, he has shown that carbon dioxide (CO_2) levels have been rising (Bates 1990; Leggett 1990; McKibben 1989). Data for 1995 show the current concentrations of CO_2 are "higher than at any time in the past 150,000 years" (Flavin 1996: 22). Increasing carbon dioxide in the atmosphere may contribute to global warming.

Scientists no longer dispute the greenhouse effect. However, some scientists (Kerr 1989; Ray and Guzzo 1992; Michaels 1992; Lindzen 1993) question the actual amount of warming that will occur and the assumptions underlying computer model projections. Skeptics suggest that the earth's natural atmospheric process (e.g., oceans, forests, sulfate aerosols) will be able to mitigate the greenhouse effect. Other skeptics argue that the current warming trend is a natural fluctuation in global temperatures rather than a result of human activities.

Despite the critics, the majority of scientific literature on the subject is mounting a very strong case that warming and climate change are beginning to occur. The twentieth century is already 0.6 degrees Celsius (1.08

degrees Fahrenheit) warmer than the nineteenth century, and according to "global climate records . . . the 10 warmest years in the past century have all occurred since 1980" (Flavin 1996: 22). The accelerated rate of warming has been attributed mostly to the activities of human beings (Bates 1990; Flavin 1996).

■ Causes and Consequences of Global Warming

The natural carbon cycle is altered by fossil fuel burning (automobiles, power plants, industry, and heating being the most common fossil fuel burning activities), which releases CO_2 into the atmosphere. The earth is constantly bombarded by solar radiation, some of which is absorbed and some of which is reflected back into space. When more solar radiation is trapped in the earth's atmosphere and less is reflected, the earth begins to warm. Gases like carbon dioxide, chlorofluorocarbons (CFCs), methane, tropospheric ozone, and nitrogen oxides trap solar radiation and cause the atmosphere to warm. Carbon dioxide accounts for roughly 50 percent of total greenhouse gases, chlorofluorocarbons 20 percent, methane 16 percent, tropospheric ozone 8 percent, and nitrogen oxides 6 percent (McKinney and Schoch 1996). As the atmosphere warms, it retains more water vapor due to evaporation. Water vapor is also a powerful greenhouse gas, because it traps long-waved solar radiation. This phenomenon has been recently demonstrated by satellite measurements from the Earth Radiation Budget Experiments (ERBE), which showed that as ocean and surface temperatures increase, more infrared radiation is trapped in the atmosphere (Leggett 1990).

Increasing temperatures set into motion various feedback loops that escalate the problem of global warming. A warming earth means that there is less ice and snow in mountain and polar regions to reflect back solar radiation. In addition, as the earth warms, large amounts of methane are released from ice, tundra, and mud in the continental shelves. More methane means more greenhouse gases to trap solar radiation, which in turn means hotter temperatures, and hotter temperatures mean more thawing, which means more methane gas—creating a vicious cycle of warming.

With global warming comes increased precipitation in some areas and drought in other areas because of increased evaporation due to the heat and changing wind patterns. Rising sea levels, longer and warmer summers, more severe storms, and more frequent forest fires are all likely outcomes of increased warming (Bates 1990; McKinney and Schoch 1996). Based on a plethora of global climate data, Flavin (1996) reports that sea levels are rising, summers are longer and warmer, serious storms have become more frequent and severe, and water shortages have become a chronic problem for eighty countries comprising 40 percent of the world's population.

Increased global warming will bring to extinction many more species before their natural time because of rapidly changing habitats (Leggett 1990; Weber 1993). The resulting storm and drought damage may also lead to possible food shortages, especially when one considers the expected increase in human population.

Who Are the Major Contributors to Global Warming?

Since the Industrial Revolution, human activity has added 170 billion tons of carbon to the atmosphere. Annual growth of global carbon emissions has been steady at about 2 percent per year (Flavin 1996). Some countries are guilty of emitting much more carbon than others, which has created problems in the international community with respect to formulating binding international strategies for regulating greenhouse gas emissions. The largest producers of carbon emissions (as measured by millions of tons) are the United States (1,371), China (835), Russia (455), Japan (299), Germany (234), India (222), the United Kingdom (153), Ukraine (125), Canada (116), and Italy (104). The United States also emits the most carbon per person (5.26 tons), followed by Kazakhstan (4.71 tons), Australia (4.19 tons), Canada (3.97 tons), and Russia (3.08 tons) (Flavin 1996).

Three of the reasons stated for such high carbon emissions in the United States, Canada, and Australia are "low energy prices, large houses and heavy use of automobiles." The growth in carbon emissions has increased each year in these three countries by over 4 percent (Flavin 1996: 31).

The fact that the United States has failed to take mandatory and legally binding steps to reduce its carbon emissions makes it difficult to persuade industrializing nations to slow their rate of carbon emissions and look to alternative technologies. Most U.S. measures to reduce the threat of global warming, as promulgated in President Clinton's climate action plan, are voluntarily based. The Natural Resources Defense Council claims that even aggressive enforcement of current strategies will not meet target reductions and, worse yet,

> in 1994 Congress approved only half the funds called for; the 1995 Congress made even more drastic cuts, and weakened appliance and lighting standards that had been enacted by the 1992 Congress. Recent trends suggest that United States carbon emissions will exceed 1990 levels by as much as 10 percent in 2000. (Flavin 1996: 32)

Hope is offered, however, by the fact that some industrialized countries such as Germany, France, Japan, Denmark, Switzerland, the Netherlands, Russia, Poland, Ukraine, and the United Kingdom have either lowered their carbon emissions or are expected to by the year 2000 (Flavin 1996).

■ International Climate Control Policies

In 1972, the United Nations (UN) Conference on the Human Environment, or the Stockholm conference, was held. This conference consisted of 114 governments and was attended by many nongovernmental organizations (NGOs). It was the first time in history that nations of the world came together to discuss issues surrounding the destruction of the environment (Switzer 1994; Valente and Valente 1995). The Stockholm conference did not create any binding obligations but served more as a catalyst to generate an international discourse on global environmental issues.

It was not until the 1992 Earth Summit (also known as the United Nations Conference on the Environment and Development, or UNCED) that serious discussion on reducing CO_2 emissions to curtail global warming was undertaken. The Rio de Janeiro conference was attended by 178 countries and 110 heads of state (Switzer 1994). One of the five major documents produced at the Earth Summit was the Framework Convention on Climate Change. Its purpose was "stabilization of greenhouse gas concentrations in the atmosphere at a level that would prevent dangerous anthropogenic interferences in the climate system" (Flavin 1996: 36).

Although the United States, headed by the Bush administration, was a major actor at the Stockholm conference, it fought binding targeted reductions in CO_2 emissions to 1990 levels by the year 2000. Despite U.S. reluctance to sign on to target CO_2 reductions, many industrialized European nations "did sign a separate declaration reaffirming their commitment to reducing their own CO_2 emissions to 1990 levels" (Gore 1992: xiv). Germany and Japan provided most of the leadership in getting the other industrial nations to make a commitment to targeted CO_2 reductions. President Clinton in 1993 reversed the Bush administration's position, announcing that the United States would reduce CO_2 to 1990 levels by the year 2000, but as noted earlier, many of the appliance and lighting efficiency initiatives enacted by the 1992 Congress have been weakened by the 1994 and 1995 Congress, which was also severely strained by receiving only half the funding needed to implement many of the voluntary programs recommended by the Clinton plan (Flavin 1996).

Getting the industrialized nations to make serious commitments has become a major concern of the Alliance of Small Island States (AOSIS) and worldwide insurance companies. The AOSIS is a small coalition of island nations that are extremely threatened by rising seas. A rise of one meter in sea level could threaten to wipe out the sustainable land and economy of many of the small island nations. Given this threat, the AOSIS proposed that industrial nations reduce their CO_2 emissions by 20 percent. This proposal was also endorsed by seventy-seven other nations participating in the Berlin conference (which was a follow-up to the Rio summit and focused on climate change) but was resisted by a majority of the oil-producing nations like Kuwait and Saudi Arabia, and by the larger carbon-consuming countries like the United States and Australia (Brown 1996; Flavin 1996).

Other major nonnation players are some of the world's largest insurance companies, which obviously have a stake in any losses that may occur due to global warming. Flavin reports that "since 1990, the worldwide insurance industry has paid out $48 billion for weather-related losses, compared with losses of $14 billion for the entire decade of the eighties" (Flavin 1996: 34).

The primary objective of the Berlin conference was to design measures that would reduce global carbon emissions and to create a series of trial projects aimed at exchanging alternative low carbon–intensive technologies between nations (Flavin 1996). Despite the fact that there were no legally binding carbon reduction targets established, the Berlin conference did provide a sense of renewed hope in formulating a global policy for mitigating climate change. The agreement reached at Berlin, known as the Berlin Mandate,

> instructs governments to negotiate a treaty protocol "to elaborate policies and measures, as well as to set quantified limitations and reduction objectives within specified time-frames such as 2005, 2010, and 2020." (Flavin 1996: 35)

This meeting took place at the end of 1997 in Japan and generated the Kyoto Protocol, which was adopted on December 11, 1997, and was opened for signatures on March 16, 1998. The protocol contains legally binding emissions targets for key greenhouse gases, especially carbon dioxide, methane, and nitrous oxide. The agreement requires ratification by fifty-five parties that must include developed countries representing at least 55 percent of the total carbon dioxide emissions of those countries. "The overall commitment adopted by developed countries in Kyoto was to reduce their emissions of greenhouse gases by some 5.2 percent below 1990 levels by a budget period of 2008 to 2012. While that percentage did not seem significant, it represented emissions levels that were about 29 percent below what they would have been in the absence of the Protocol" (UN 1998).

While continued negotiations offer hope, there is still much that has to be accomplished with respect to establishing firm, binding policies on fossil fuel reduction. Some of the options under consideration include eliminating subsidies for fossil fuel extraction and use and creating different types of energy taxes. Also under consideration are suggestions for how industrialized nations should provide financial assistance and technical support for industrializing nations. There is hope on the horizon, but until reduction ceilings for greenhouse gases can be mutually agreed upon and made legally binding, it is unlikely that much progress will be made in reducing greenhouse gases.

■ THE THREAT OF OZONE DEPLETION

Record low ozone levels have been recorded throughout the 1990s. As we will see, international policies appear to have reduced the sources of ozone depletion, but the damaging effects of a thinning ozone layer have yet to be fully realized, and it may be as long as a century before ozone levels return to concentrations prior to the creation of CFCs.

■ Causes and Consequences of Ozone Depletion

There are two types of ozone we often hear about. Ozone in the stratosphere is what protects us from harmful ultraviolet radiation. If compressed, the ozone layer would be about a tenth of an inch thick (3 mm). Ozone in the troposphere, on the other hand, is extremely poisonous to most life forms. A large portion of the ozone found in our troposphere is generated by human sources of atmospheric pollution that interact with solar radiation to create a pale blue gas with a strong pungent order, which can sometimes be smelled after it rains (Gribbin 1988). The issue we are concerned with in this chapter is the depletion of protective, stratospheric ozone resulting from the human production of chemicals like CFCs, halons, and other halon carbons. This stratospheric ozone layer protects us from excessive amounts of ultraviolet radiation.

The ozone layer is created from oxygen that escapes from the troposphere. Oxygen is created from living organisms that process carbon dioxide and exhale oxygen. Just as oxygen is vital for the creation of the ozone layer, so is the ozone layer vital for the creation of oxygen by protecting the living organisms that produce it.

CFCs and halon molecules, which have relatively long atmospheric lives, contain chlorine and bromine atoms. Once these molecules float their way up into the atmosphere, they interact with sunlight, which breaks apart their molecular structure releasing the chlorine and bromine atoms that destroy ozone. It is known that "a single chlorine atom can scavenge and destroy many thousands of ozone molecules" (Gribbin 1988: 48). Gribbin explains:

> At an altitude of about 11 miles above the ground, more than half of the ozone above Antarctica was destroyed in the spring of 1987. And changes in the amount of chlorine oxide present marched precisely in step with changes in the amount of ozone. Where chlorine oxide went up, ozone went down, showing clearly that chlorine was destroying the ozone. (Gribbin 1988: xi)

The chlorine atoms that destroy ozone are a product of chemical reactions caused by high-energy ultraviolet radiation. Ultraviolet radiation breaks apart the CFC molecule in the stratosphere separating the chlorine atoms, which then destroy large amounts of ozone molecules.

CFCs are solely products of human industry and are most often used for propellants in aerosol spray cans. They are also used in air conditioners, refrigerators, computer chips, and styrofoam. Halons are used mostly in equipment used to suppress fires. There are no natural processes in the troposphere that react with CFCs and halons to break their molecular structure down (McKinney and Schoch 1996).

Another major destroyer of ozone is nitrous oxide (N_2O). Like CFCs and halons, N_2O is not a friend to the stratosphere. Nitrous oxide is emitted into the atmosphere by plants, combustion of coal and oil, and spray cans. Plants naturally emit N_2O, which helps maintain ozone levels so as to strike a balance between too little ozone—a certain amount is needed to protect the earth from harmful radiation—and too much ozone, which would inhibit life as we know it (Gribbin 1988; McKinney and Schoch 1996). While the release of N_2O by plants is a natural process, the use of chemical fertilizer to increase food production has increased the amount of N_2O in the troposphere (Gribbin 1988).

Effects of ozone depletion on human beings are several. The most obvious is that ozone depletion causes an increase in ultraviolet B (UV-B) radiation, which is known to increase our chances of getting skin cancer, especially malignant melanoma (Meadows, Meadows, and Rander 1992). Increases in UV-B radiation have been most pronounced in countries located in the Southern Hemisphere, such as Australia, New Zealand, and South Africa. Australia has the highest rate of skin cancer in the world, and researchers suggest that two out of three people growing up there will develop skin cancer during their lifetimes, "and 1 in 60 will develop the most deadly type, malignant melanoma" (Meadows, Meadows, and Rander 1992: 145). In the Southern Hemisphere, skin cancer and cataracts are increasingly common. In fact, Al Gore reports that

> in Queensland, in northeastern Australia, for example, more than 75 percent of all its citizens who have reached the age of sixty-five now have some form of skin cancer, and children are required by law to wear large hats and neck scarves to and from school to protect against ultraviolet radiation. (Gore 1992: 85)

In addition to increased rates of skin cancer, UV-B radiation also suppresses the human immune system, making us much more vulnerable to disease and viruses (Gore 1992; Gribbin 1988; Meadows, Meadows, and Rander 1992; McKinney and Schoch 1996).

Increased UV-B radiation on plants such as soybeans, beans, sugar beets, potatoes, lettuce, tomatoes, sorghum, peas, and wheat has been shown to inhibit growth, photosynthesis, and metabolism (McKinney and Schoch 1996). High levels of UV-B radiation also affect freshwater and marine ecosystems, especially ocean plankton, the base of the ocean's food chain (McKinney and Schoch 1996). There have been reports from Chile that sheep are going blind and that rabbits are developing myopia so severe that one can walk into a field and pick them up by the ears (Lamar 1991). Cattle are also known to get eye cancer and pinkeye when exposed to high levels of UV-B radiation (Gribbin 1988).

■ Major Producers and Users of Ozone-Depleting Substances

CFCs, like many other chemicals, really became popular after World War II. The production of the two most common types of CFCs increased from 55,000 tons in 1950 to 800,000 tons in 1976:

> From 1950 to 1975 world production of CFCs grew at 7% to 10% per year—doubling every 10 years or less. By the 1980s the world was manufacturing a million tonnes of CFCs annually. In the United States alone CFC coolants were at work in 100 million refrigerators, 30 million freezers, 45 million home air conditioners, 90 million car air conditioners, and hundreds of thousands of coolers in restaurants, supermarkets, and refrigerated trucks. (Meadows, Meadows, and Rander 1992: 142)

Approximately 90 percent of the CFCs used were immediately released into the atmosphere during their use, with the other 10 percent being released after the product was discarded (e.g., refrigerators and air conditioners) (McKinney and Schoch 1996).

Like most pollution caused by modern technology, there is an immense disparity between the amount of ecological destruction caused by industrialized and industrializing countries. The United States, the European Community (now the European Union), Japan, New Zealand, and Australia have used much more CFCs than industrializing countries like China and India (USCC and AN 1991). When broken down by per capita use, North Americans and Europeans were using on average 2 pounds (0.85 kg) of CFCs a year per person, and developing countries like China and India were using less than an ounce (0.03 kg) a year per person (Meadows, Meadows, and Rander 1992).

In 1985, there was no longer any doubt that CFCs were causing the destruction of the ozone layer. When British scientists measured a 40 percent decrease in ozone over Halley Bay in Antarctica, it was perceived as an error. After checking their measurement and other monitoring sites, it became apparent that ozone depletion was happening. NASA confirmed with readings made by the Nimbus 7 satellite that a hole in the ozone layer was indeed open over Antarctica. This discovery prompted a series of international meetings that led to a major conference designed to create target reductions and phaseouts of ozone-depleting substances.

■ International Policies on Ozone Depletion

Sherwood Roland and Mario Molina's 1974 paper documenting the depletion of ozone led to policies both in the European Community and the United States requiring the phaseout of spray cans that used CFCs as propellants. In 1978, the United States banned the use of CFC propellants in spray cans, and the European Community reached a voluntary agreement requiring a 30 percent reduction in CFC propellant spray cans (Gribbin 1988; USSCEPW 1993). Despite the fact there were no international agreements at that time, there was a decrease in CFC propellants between 1974 and 1982 in most of the developed world except the Soviet Union (Gribbin 1988). The decrease in CFC propellants for spray cans, however, did not include a decrease in other uses and types of CFCs.

In 1985, the same year the ozone hole over Antarctica was discovered, an international agreement, titled the Vienna Convention for the Protection of the Ozone Layer, was signed by twenty nation producers of halocarbons. The agreement, however, did not entail any binding phaseout or reduction of ozone-depleting substances. Instead, the agreement focused on international research efforts aimed at documenting the ozone-depleting potential of halocarbons and other CFC and non-CFC ozone-depleting substances (Gribbin 1988; USSCEPW 1993).

After NASA's confirmation of the Antarctic ozone hole in 1987, it took only nine months of negotiations before twenty-seven countries signed the 1987 Montreal Protocol on Substances that Deplete the Ozone Layer. The Montreal Protocol is by far the strongest piece of international environmental policy to date. The agreement became effective January 1989 and required a "freeze in worldwide production of CFCs and halons (at 1986 levels, for CFCs in 1989 and halons in 1992) and a 50 percent reduction in the production and consumption of CFCs by mid-1988" (USSCEPW 1993: 108). The twenty-seven nations signing this document accounted for 99 percent of the producers and 90 percent of the consumers of ozone-depleting substances. Declining to sign, however, were many industrializing nations like India and China, because the original agreement made no stipulation for providing technical and financial assistance for developing nations (USSCEPW 1993).

Due to declining record ozone levels over the Northern Hemisphere reported by the scientific community in 1989 and again in 1992, the Montreal Protocol has been amended twice since its original signing. The first amendment took place in June 1990 in London. The amendment included an agreement by all the parties "to a complete phaseout of CFCs, halons and carbon tetrachloride by the year 2000" (USSCEPW 1993: 110). Further agreements included a ban in the year 2005 on the production of methyl chloroform. The London amendment also generated an agreement between the original Montreal Protocol signers to create a fund to provide assistance to developing nations for converting to the use of ozone-friendly substances (USSCEPW 1993).

A second major amendment was called for after the scientific community reported more record losses in ozone over the Northern Hemisphere (USSCEPW 1993). The second amendment was signed in November 1992 in Copenhagen by 126 countries. The amendment established more rapid phaseout dates for the major ozone-depleting substances. In addition to the phaseout of CFCs, hydrochlorofluorocarbons (HCFCs), the substance used to replace CFCs, are also targeted for phaseouts beginning in 2004. While HCFCs are much less devastating to the ozone layer, they are known to reach the stratosphere with 2 to 5 percent of the ozone-destroying potential (McKinney and Schoch 1996). The full ramifications of HCFCs to the atmosphere and human health are not yet known, but the fact that the Copenhagen agreements call for a total ban on HCFCs by the year 2020 ensures that their overall impact on the environment will be mitigated.

The Montreal Protocol and its amendments are testimony to the positive environmental policy that can be promulgated among nations when environmental degradation is taken seriously. It shows us that serious environmental degradation has a way of smoothing over ideological differences between nations. The Montreal Protocol and its amendments were passed because a diverse group of international scientists, politicians, and corporations agreed that preservation of the ozone layer—a necessity for life as we know it—outweighed the subtle political and economic differences that most often keep nations from international agreements. The Montreal Protocol and its amendments are the most successful international environmental policy to date.

■ THE ACID RAIN PROBLEM

Acid rain (often referred to as transboundary air pollution or simply air pollution) has become a major problem throughout the world, but it is most publicized in the United States, Canada, and Europe. Acid rain kills lakes and rivers, seriously damages the soil, and endangers the health of animals and humans. Like most pollution, acid rain does not stop at the political border of one country and ask permission to enter another. Acid rain has become a source of conflict between the United States and Canada and among many European nations.

■ Causes and Consequences of Acid Rain

Acid rain is created almost immediately after sulfur dioxide has been emitted into the atmosphere. Almost all fossil fuels contain sulfur, and sulfur content is extremely high in coal. In the United States, coal-fueled utility power plants account for roughly two-thirds of the U.S. sulfur dioxide emissions (Switzer 1994). When fossil fuels burn, sulfur combines with oxygen to create sulfur dioxide (SO_2), which is an odorless and colorless gas. Sulfur dioxide is a known lung irritant that can, in low concentrations, bring about asthmatic attacks and make those with respiratory problems quite uncomfortable. In the United States, sulfur dioxide by itself has not been a major problem, but the transformation of sulfur dioxide in the atmosphere into sulfuric acid (H_2SO_4) is a major environmental health problem (Seis 1996).

In the atmosphere, SO_2 combines with oxygen to form sulfate (SO_4). Sulfate is a small particle that floats in the air or settles on leaves, buildings, and the ground. When sulfate interacts with mist, fog, or rain it becomes acid rain (USCC and AN 1991). Sulfate, when inhaled into the moist lungs, is also transformed into sulfuric acid. All told, acid rain is dangerous to human health and to ecosystems in general, especially to trees, rivers, and lakes.

The disappearance of fish in the northern United States and in Canada has been on the increase since the early 1970s. According to one 1984 report, "In the Adirondack mountains, at least 180 former brook trout ponds will no longer support populations" (USCC and AN 1991: 3652). In 1975, another study, which surveyed 214 Adirondack lakes, showed that "90 percent of these lakes were entirely devoid of fish life" (USCC and AN 1991: 3652). The Office of Technology Assessment "estimated that in the Eastern United States approximately 3,000 lakes and 23,000 miles of streams have already become acidified or have virtually no acid neutralizing capacity left" (USCC and AN 1991: 3655). In Europe, "fish have disappeared from lakes in Sweden and Norway, as well as Scotland and England" (Switzer 1994: 258). Lake fish populations have been declining and in some cases dying out all together in countries like Russia and Romania (Switzer 1994).

Acid rain has also been linked to forest declines in various parts of the United States. In Eastern Europe forests are dying rapidly, and in southwestern Poland, the army was used in 1990 to fell large tracts of dead forest due to acid rain. Military factories pumping out large doses of sulfur dioxide along the Russia-Finland border are responsible for ravaging forests within a 300-mile radius of the factories and are responsible for damage to "an additional fifty thousand square miles, with an estimated 30 percent of the firs in Finnish Lapland in danger of dying" (Switzer 1994).

The effects of acid rain on human health are beginning to make an appearance in populations throughout the world. In Santa Catarina, Brazil, for example, "the environmental secretary estimates that 80 percent of the local hospital patients have respiratory ailments caused by acidic pollutants" (Switzer 1994: 265). High acid levels have been correlated with increased colds, bronchial infections, asthma attacks, and death. Harvard public health researchers suggest that approximately a 5 percent annual excess of mortality in the United States is due to sulfate and fine particles (USCC and AN 1991).

■ Major Acid Rain Producers

As in the cases of ozone depletion and global warming, it is the largest industrialized nations that are the major generators of acid rain–causing pollutants. Coal-fueled power plants and factories account for the highest emissions of sulfur dioxide leading to acid rain worldwide. Thirty-nine percent of worldwide electricity is generated from coal (Flavin and Lenssen 1991). In the United States alone, 75 percent of total sulfur emissions come from power plants and large factories, which also emit large amounts of oxides of nitrogen (Switzer 1994).

The biggest problem with the countries producing the most sulfur dioxide and oxides of nitrogen emissions is that their pollution becomes the problem of other countries. Air pollution goes where the wind blows. Most of Canada's acid rain problems have been attributed to coal-fueled power plants operating in the Ohio Valley. Likewise, coal-fueled power plants and factories in eastern Germany, Poland, the Czech Republic, and the Slovak Republic destroy forests to the east. Sulfur emissions from Russia have been implicated in an acid rain problem for Finland and Sweden, and emissions from Great Britain have contributed to acid rain problems for Norway. Sulfur emissions from China create acid rain problems for Japan and South Korea.

The acid rain problem is severe in Europe; the International Institute for Applied Systems Analysis has estimated that "75 percent of Europe's forests are now experiencing damaging levels of sulfur deposition" (Brown 1993: 6). The monetary damage assessment for Europe is estimated to be at $30.4 billion per year, which does not take into consideration the other ecological functions forests serve with respect to regulating climate, flooding, and erosion (Brown 1993).

Given the fact that coal-powered electricity is the major culprit in acid rain generation, reducing our reliance on coal to generate electricity seems to be the obvious answer. No solutions between nations, however, are easy. Coal is a cheap and widely available fossil fuel when compared with more expensive energy sources (e.g., hydroelectric power, natural gas, and oil) and in some cases more dangerous alternatives (e.g., nuclear power). Thus, solving what we know to be a simple problem becomes extremely complicated when we figure into the solution domestic and international politics and economics.

■ International Policies on Acid Rain

There have been no sponsored United Nations multilateral agreements regarding the abatement of acid rain. Unlike global warming and ozone depletion, which cause worldwide problems, acid rain is more regional. Accordingly, most acid rain agreements tend to be signed between bordering nations.

Probably the most extensive acid rain discussions have been between the United States and Canada. Canadian environmentalists contend that sulfur dioxide emissions from the United States are responsible for damaging as many as 16,000 Canadian lakes (Switzer 1994). Serious talks between Canada and the United States began in 1978 and culminated in a nonbinding treaty titled the Great Lakes Water Quality Agreement. The treaty stipulated measures that required emission reductions on the part of both nations. The progress made in 1978 was negated in 1980 when President Carter, acting on the Middle East oil crisis, decided to convert 100 oil-fired utility plants to coal (Switzer 1994).

From 1980 to 1990, acid rain talks between the United States and Canada did not make much progress. It was not until 1990, when President Bush signed the 1990 Clean Air Act (CAA), that serious efforts were initiated to reduce U.S. sulfur dioxide and nitrogen emissions.

Europe has also been struggling with international strategies to curtail acid rain. The European Union (EU) and the UN Economic Commission for Europe (ECE) have been instrumental in at least initiating efforts to abate acid rain. The EU consists of fifteen member nations, and the ECE consists of all the nations of Europe. In an effort to make economic and environmental laws uniform, the EU has experienced difficulty in attempting to promulgate uniform sulfur dioxide emission reductions across nations. In fact, Great Britain, the largest producer of sulfur dioxide emissions in Western Europe, has often balked at emission standards despite being a country that has suffered from air pollution disasters (Switzer 1994).

The ECE has made more progress, beginning in 1979 with the enactment of the Convention on Long-Range Transboundary Air Pollution. This agreement of intent obliged each nation to develop technology to abate acid rain and where appropriate to share the technology. The agreement, however, established no emissions reduction standards nor did it require any uniformity in implementation of acid rain reduction technology. It was not until 1985 in Helsinki that thirty European nations agreed to a 30 percent reduction in sulfur dioxide emissions by the year 1993. Those that signed the protocol are known as members of the "30 percent club." While most European nations signed on, Great Britain, Spain, Ireland, Greece, and Portugal did not. Unfortunately, the 30 percent reduction is a politically derived target, not a scientifically based one, which means that there are no guarantees that a 30 percent reduction will abate the effects of acid deposition throughout Europe (Switzer 1994).

Countries like Norway, Sweden, Finland, and Japan have been making major inroads into generating sulfur dioxide–reducing technology and abating their own sulfur dioxide emissions. Due to the effects of acid rain on their own forests, Norway, Sweden, and Finland made Russia a $1 billion loan to utilize Finnish desulfurization technology (Switzer 1994). These same countries have been extremely influential in convincing most of Europe of the seriousness of acid rain. Japan has also been instrumental in helping the Chinese develop desulfurization technology. Unfortunately, many environmentalists contend that good desulfurization technology may not be enough for China, which has huge coal reserves and is just beginning to accelerate toward industrialized development (Brown 1993, 1996; Leggett 1990; Postel 1994).

Many nations are beginning to recognize the damaging effects of acid rain and are beginning to act by establishing sulfur dioxide emission reductions. The acid rain problem, however, is complicated for many reasons. Coal is the cheapest and most abundant fossil fuel remaining, and for developing nations it is at this time the only economically viable option for pursuing development. Even many of the more developed nations are reluctant to abandon coal as a major fuel. Although some reduction strategies are being implemented, it seems apparent that acid rain is not going to diminish as a major environmental problem anytime soon.

■ CONCLUSION

Of the two major climate problems we have discussed, global warming appears on the international level to rank as less serious than ozone depletion. The major reason global warming appears to be a subordinate concern is probably due to the uncertainty that surrounds the possible outcomes of a warming planet. Furthermore, the causes of global warming are inextricably intertwined with politics, economics, growth, and development. Because global warming is taken less seriously, as judged by the commitment and legal teeth of international agreements, it could prove to be the world's most serious future problem. Radical climate change and climate unpredictability could create worldwide food and housing shortages. On a more optimistic note, however, current international efforts toward reducing greenhouse gases are much more serious than they were at the Rio summit.

International policy regarding ozone depletion is ecologically sound and moving toward a complete worldwide phaseout of major ozone-depleting substances. Unfortunately, it is going to be decades before international agreements produce major reductions in ozone depletion and a few decades more before ozone restoration becomes discernible. Nevertheless, the Montreal Protocol and its two amendments epitomize the type of international environmental agreements that can be achieved when nation-states recognize how they are interconnected ecologically.

Acid rain is also forcing nation-states to see beyond their political boundaries. Some countries are making major efforts to reduce emissions responsible for acid rain, but many nations are not, because they are rich in low-grade coal and poor in desulfurization technology. Unfortunately, from an ecological standpoint, low-grade coal is the most available fossil fuel reserve for many countries, which means that acid rain will most likely continue to be a bioregional and geopolitical problem. The best hope for acid rain reduction worldwide lies in emission reductions of low-sulfur coal for industrialized nations and the easy availability of high-tech desulfurization equipment for all industrializing nations.

As we have seen in this chapter, atmospheric pollution influences the climate, and changing climate affects the fertility and integrity of the land. The next chapter examines ways in which atmospheric pollution is directly linked to land resources.

■ QUESTIONS

1. What are some of the ways the largest greenhouse contributor nations could reduce their emissions?

2. What are some of the major differences between the ozone problem and the global warming problem? Why is it so difficult for the world community to reach a viable solution regarding global warming, like they did with ozone depletion?

3. In what ways is acid rain an international problem? Describe some solutions to the acid rain problem if nations worked together.

4. Does the North have an obligation to help the South develop in a more environmentally safe way? Does it have an interest in helping?

■ **SUGGESTED READINGS**

Bates, A. K. (1990) *Climate in Crisis*. Summertown, TN: Book Publishing Company.

Brown, Lester R. (1996) *State of the World*. New York: W.W. Norton.

Brown, Lester R., N. Lenssen, and H. Kane (1995) *Vital Signs 1995: The Trends That Are Shaping Our Future*. New York: W.W. Norton.

Commoner, Barry (1990) *Making Peace with the Planet*. New York: Pantheon Books.

Gleason, J. F., et al. (1993) "Record Low Global Ozone in 1992," *Science* 260, no. 5107.

Gribbin, J. (1988) *The Hole in the Sky: Man's Threat to the Ozone Layer*. New York: Bantam Books.

Leggett, J. (1990) "The Nature of the Greenhouse Threat." In J. Leggett, ed. *Global Warming: The Greenpeace Report*. New York: Oxford University Press.

Meadows D. H., D. L. Meadows, and J. Rander (1992) *Beyond the Limits: Confronting Global Collapse, Envisioning a Sustainable Future*. Mills, VT: Chelsea Green Publishers.

Rifkin, J. (1989) *Entropy: Into the Greenhouse World*. New York: Bantam Books.

Schnaiberg, A., and K. A. Gould (1994) *Environment and Society: The Enduring Conflict*. New York: St. Martin's Press.

Switzer, J. V. (1994) *Environmental Politics: Domestic and Global Dimensions*. New York: St. Martin's Press.

Cooperation and Conflict over Natural Resources

Karrin Scapple

Wars have been fought and peace has been waged over natural resources. States rely on natural resources to sustain their economies and their sense of independence. Since few states are self-sufficient, they often cooperate with other countries to obtain natural resources that they need; if cooperation is not possible, violent conflict becomes a viable alternative. The dilemma for a leader is to determine how to fulfill the state's needs with as little conflict as possible and without relinquishing too much sovereignty (state sovereignty). The resolution to this dilemma often depends on which natural resource is involved.

■ WHAT ARE NATURAL RESOURCES?

There are many types of natural resources and they have different impacts on global politics. Natural resources can be identified by whether they are renewable or nonrenewable, and whether they stay within a single border or are transboundary.

■ Renewable Versus Nonrenewable Resources

A renewable resource is one that regenerates itself, such as trees, fish, and animals. Conversely, a nonrenewable resource is one that does not regenerate; once it is used, it cannot be recreated. In many respects *nonrenewable resource* is a misnomer. Most resources are, in fact, renewable. The issue becomes whether the resource can be renewed over a reasonable period of time in human terms. For instance, oil is a renewable resource and can regenerate over time; however, we would need to measure the regeneration time in centuries, rather than months. So for policymaking purposes, it is more accurate to consider oil, and other fossil fuels, as a *nonrenewable* resource.

Renewable and nonrenewable resources have different impacts on the international system. Theoretically, states should not have to fight over renewable resources because of their regenerative characteristic. If states are unable to meet their own needs for a particular resource, they are likely to cooperate to meet those needs through trade agreements and/or economic integration. Conflict results, however, if the needed renewable resource is overconsumed so that full regeneration, or sustainable growth, is no longer possible. Fishing and whaling conflicts have become key issues over the past several decades.

Conflict and violence sometimes result over nonrenewable resources. If a resource is needed but there is a finite amount available, states will sometimes fight very hard to obtain that resource. The Gulf War is an example of a highly visible conflict over nonrenewable resources. While cooperation is possible, it becomes less likely if the nonrenewable resource is, like water, critically needed and if the disputants have unresolved conflicts from the past.

■ Boundary Versus Transboundary Resources

The question that must be addressed here is whether the resource stays in one place or whether it moves. A forest is an example of a boundary resource; it is located within a state's borders and ownership is clear. Conflict is less likely with a boundary resource because of the international principle of sovereignty. A river, however, is an example of a transboundary resource that may either define a border between two countries or travel from one country to another; in either case, the river must be shared by two or more states. Although the opportunity for cooperation increases in a transboundary resource, so does the possibility for conflict.

The issue of sovereignty becomes critical when one looks at natural resources in boundary and transboundary terms. International law protects a state's sovereignty and its territorial integrity. International law asserts that the resources found within a state's borders are that state's property to do with as it chooses. Although countries have concerns over Nigeria's human rights abuses involved in oil development, no country disputes Nigeria's right to develop its oil fields. The oil fields are in Nigerian territory and are under Nigerian sovereignty.

Yet, this sense of sovereignty becomes controversial with transboundary resources. If a river head exists in one state, does that state have the right to do whatever it wants with the water, even though downstream countries may be dependent on that resource too? The answer to this question depends on whether a country is "upstream" or "downstream." Upstream states tend to rely heavily on the sovereignty principle; downstream states tend to promote the idea of sharing and cooperation. Cooperation may be relatively easy over some resources, but conflict becomes more likely if the resource is critical to survival.

■ CASE STUDIES

Using the factors of renewability and location, a fourfold matrix can be created (see Table 14.1). The remainder of this chapter will focus on a case study in each of the cells of this matrix so that some of the issues of natural resources can be highlighted. These cases provide examples of how natural resources can be a center of cooperation as well as a source of conflict in the international community.

▨ Renewable, Boundary Resource: Forests

It has already been noted that the use of a boundary resource, such as a forest, is clearly within the domain of the sovereign state because the resource does not travel outside its borders. While this is true, use of a boundary resource can often have second-order consequences for surrounding states or for the global community at large. For instance, while a forest does not travel, a river that runs through a forest does; if that forest is clear-cut, there will be enormous ramifications on the river as a result of sediment increase or flooding. Further, that forest's contribution to regional and global climate conditions will be lost, with possible serious implications. (The role of forests in global climate change is discussed in the next chapter.) As a result, the international community has come to be concerned about a state's use of its resources when there are possible global consequences.

There are debates about who should have jurisdiction over a boundary resource when there are second-order consequences. The North is currently advocating that the needs of the international community should be

Table 14.1 Natural Resource Matrix

	Renewable	Nonrenewable
Boundary	Forests	Oil
Transboundary	Fish	Water

a priority. Issues such as global climate change, stratospheric ozone depletion, and law of the sea should take precedence in policymaking decisions. The South, however, is primarily concerned about its own development opportunities. Boundary resources may provide a good source of export income, as well as meet internal development needs. These states believe they have the right to develop and meet the needs of their people. Further, the South believes that environmental concerns are being used as yet another way to ensure that the South does not develop and fully compete with the industrialized North.

The irony is that not only do many of the development patterns in the South actually limit future potential for growth, but the North makes a great contribution to habitat destruction. The nutrients in most tropical forests are found in the vegetation, not the soil; a forest that is clear-cut can be productive for only three to five years. Thus, it may be in the South's best interests to find other ways of developing, rather than through deforestation. Also, the North has some culpability in this process. U.S. and European demands for products made from mahogany and teak, as well as the penchant for inexpensive beef, has created the trade environment that encourages deforestation.

The dilemma is to determine how forests can be used for development and at the same time continue to make their contributions to the ecosystem. One solution that has met with some success has been debt-for-nature swaps.

Debt-for-nature swaps in Costa Rica. Cooperation is the key word in debt-for-nature swaps. Several actors are involved: financial institutions, nongovernmental organizations (NGOs), and states. These actors negotiate an arrangement that reduces the state's international debt while also preserving some of its forests for conservation and sustainable development. Most states that are heavily deforesting are also heavily in debt; Costa Rica is one such state. While each case differs, the general arrangement is that an NGO in the North (Conservation International, World Wildlife Fund, and Nature Conservancy have been the most prominent) raises funds to buy-down a portion of the state's international debt at a reduced rate (see Chapter 7). In exchange, the government agrees to give a local NGO the equivalent amount of money in local currency (or issues a bond) to create a park out of what would have become deforested land. It appears to be a win-win situation: the financial institution gets something for a debt that may not otherwise be repaid; the state gets part of its international debt eliminated; and the NGOs help save part of a natural forest that can continue to contribute to the global climate system.

More than 25 percent of Costa Rica, a country about the size of West Virginia, consists of parks and protected areas, yet it also has one of the highest deforestation rates in the world, at almost 3–4 percent annually (Jones 1992; Sarkar and Ebbs 1992). Deforestation has occurred because of the combined pressure of population growth and industrialization. The strain of development, combined with a diffuse bureaucratic system, has caused many of the parks to be undermanaged by government agencies and penetrated by individuals in search of meeting basic human needs. Even though there have been legal efforts to curb deforestation since the mid-1800s, enforcement remains a problem (Jones 1992). One of the critical needs is money to help protect the existing park system.

Several years ago, it was estimated that Costa Rica would have to use 25 percent of its annual budget to service its international debt (Sarkar and Ebbs 1992). Clearly, in this situation, supporting a forest would be considered a "luxury." Costa Rica found that debt-for-nature swaps were a way to resolve the problem, and it has been the most active state in terms of the number of swaps, the amount of money generated, and the amount of debt reduced (Mahony 1992).

There are many benefits of debt-for-nature swaps. They provide financial support for environmental and resource issues that would otherwise be unavailable. International debt is reduced, and there is cooperation to resolve the problem between the North and South; active participation from local NGOs is a possibility as well.

There are also disadvantages. Many banks are unwilling to sell the debt at a reduced rate. In fact, the World Bank and other multilateral development banks are prohibited from doing so; thus, about 60 percent of the global debt is beyond the reach of the debt-for-nature process (Klinger 1994). Even with the eligible debt, many critics are concerned that only a small amount of the debt is actually eliminated (Mahony 1992). Costa Rica, the model for the debt-for-nature concept, has been able to eliminate only 5 percent of its international debt (Sarkar and Ebbs 1992). There are continuing concerns that the parks created are still not well managed, and deforestation persists despite the transaction. Finally, some argue that it is still the North that benefits most from the arrangement and that it has domination over the process (Mahony 1992).

Despite the controversy, debt-for-nature swaps offer a unique way to resolve natural resource issues. They provide an opportunity for cooperation at many levels: between North and South, between bank and debtor, and between governments and nongovernmental organizations. With some modifications to the system, debt-for-nature swaps might be a valuable model for resolving conflict over other natural resource issues.

■ Renewable, Transboundary Resource: Fish

Any child who has owned a couple of guppies can tell you that fish are a renewable resource. In fact, fish renew at a very fast rate. Yet, according to the UN Food and Agriculture Organization, catches are declining in almost 50 percent of the global marine fish stocks (WRI 1994). Nine of the seventeen major fishing grounds have been devastated from overfishing, and four more are seriously threatened. Yet, over one-fifth of the global population rely on fish as their main source of protein (Ghazi, Smith, and Trevena 1995). As the population increases, there will be more demands for inexpensive food sources, such as fish.

One of the problems is that as fishing areas close to coastlines are depleted, people have to travel farther to find a reasonable catch. These travels often take fishermen beyond their territories and into waters that other states believe are within their jurisdiction. In addition, fishing companies have created new technology to increase their catch size. They use drift nets that allow them to "fish" over a 20–40-mile span, and they employ factory ships that allow them to more efficiently process the fish right on the ships. The combination of these new technologies, coupled with the growing demand for ocean resources, has led to several conflicts over fishing rights.

The United Nations (UN) led the way in resolving these conflicts by initiating negotiations to create a treaty that would help manage ocean resources. The UN Conference on the Law of the Sea involved a series of meetings, from 1973 to 1982, that culminated in the signing of the Law of the Sea Treaty (LOS). One of the major results of LOS was that territorial waters were given political definitions. According to LOS, a state has jurisdiction over its territorial waters, which are defined to be 12 miles from the coastline; this is an area that the state may defend as part of its territory. Beyond that point, up to 200 miles, the area is considered an exclusive economic zone (EEZ); this area is not part of the sovereign state, but the state has the exclusive right to control the ocean resources in the area. This means that the state can use all of the resources found within the EEZ or sell rights to other states to use the resources. Anything beyond the EEZ is considered global commons area, an area that is to be shared by all peoples and to be managed by the international community.

While there are still many problems with managing the resources in the global commons area and resource depletion is a major concern, it was envisioned that LOS would prevent most of the conflict over fishing rights. Coastal states would no longer be threatened by foreign fishing ships right off their coasts; the limits of the coastal states' jurisdiction were now clearly determined, and the coastal states had the opportunity to sell rights to use resources in the EEZ. It was believed that through multilateral cooperation, violent conflicts over this limited resource would be avoided.

Fish wars: The fight for turbot. Violence became imminent between Canada and the European Union (EU) over fishing rights in the Grand Banks, more than 200 miles off the coast of Newfoundland, Canada. Even though the fishing grounds were beyond its EEZ, Canada believed it had the jurisdiction to protect the "straddling" stock of turbot. A straddling stock consists of fish that literally straddle a state's EEZ and the global commons area. According to international law, a state has jurisdiction over the stock when it is in the EEZ, but does not have jurisdiction when the stock is in the high seas. The Northwest Atlantic Fisheries Organization (NAFO), an international governmental organization created to manage the fishery, set quotas for 1995: 3,400 tons for EU boats and 16,300 tons for Canadian boats. Canada said that trawlers from both Spain and Portugal (both members of the EU) had been fishing over the NAFO-set quotas for the previous five years; someone needed to take action to protect the stock, and Canada believed it had the right to do so (DeMont 1995a). On March 3, 1995, Canada declared a sixty-day moratorium against turbot fishing while the parties could negotiate an agreement over the quotas. When Spain and Portugal refused to honor the moratorium, the situation grew tense.

The Canadian coast guard attempted to board one of the Spanish fishing trawlers, the *Estai*. When their attempts were evaded, the Canadians fired warning shots over the bow of the *Estai*, seized it, and jailed the crew. Canada also cut the nets of several other boats. When the Canadians examined the nets, they found that the mesh was too small and that only

Source: Danziger, *The Christian Science Monitor,* March 29, 1995. © *The Christian Science Monitor.*

2 percent of the turbot catch had reached spawning age; both actions violated international agreements (Russell 1995). EU officials threatened to send warships. Spain threatened to take the case to the International Court of Justice and sent several navy patrol vessels to the Grand Banks area.

After many weeks of threats and negotiations, an agreement was finally reached. Canada gave up some of its catch quota to the EU in exchange for the implementation of tough surveillance and enforcement measures, including on-board inspections and satellite monitoring (De-Mont 1995b). Cooperation ultimately prevailed, but it took the threat of violence, from a usually quiet country such as Canada, to initiate an agreement. Yet, this agreement will be more likely to protect the turbot stock than the previous arrangement.

■ Nonrenewable, Transboundary Resource: Water

The Quran states that water is the source of all life, but water is the primary limiting resource in an arid region. In fact, some experts believe that the next protracted war in the Middle East will be fought over water, not oil (Gleick 1994; Postel 1993). These states depend almost entirely on river systems for their water supply. The problem is that none of the rivers exists solely within one state's borders; all are transboundary. Generally, countries that are in control of the water supply are in the most powerful position, since they can control both the quantity of water (through dams) and the quality of water (by their industrial and agricultural actions). This creates a power imbalance between upstream and downstream states, which can lead to either cooperation or conflict.

One of the debates concerns the issue of sovereignty. A guiding principle in international law states that a government has jurisdiction over domestic issues; but the world is becoming more interdependent each day. The boundary between domestic and international issues is rarely clear, as has been noted in many chapters in this book.

Proponents of absolute sovereignty rely on the Harmon doctrine: a state has the right to make all decisions about the resources that lie within that state's territory. The growing awareness of transboundary resource problems has led to the development of an alternative doctrine—equitable utilization—promoted by the United Nations: states have the obligation to ensure that their use of those resources will not adversely affect other states (Ahmed 1994). There is still great debate as to which principle should prevail in resource issues. This is particularly true in a region, like the Middle East, that is highly dependent on a transboundary resource for its very survival.

The Tigris-Euphrates river system. The relationship between Turkey, Syria, and Iraq over the Tigris and Euphrates rivers is a good example of the typical upstream/downstream relationship that can be characterized by conflict rather than cooperation. These rivers begin in the Turkish mountains, diverge before they reach Syria, and then join again farther downstream in Iraq. Approximately 40 percent of the water resides within Turkish territory, while 25 percent is in Syria and 35 percent is in Iraq (Hillel 1994). All three states are in the process of developing, and their success in this area will be dependent in large part on their ability to maintain a consistent and adequate water supply.

Turkey has taken the greatest steps to this end by creating an elaborate dam system (the Southeast Anatolia Project) to create storage, generate hydroelectric power, and prevent floods. When completed, the project will involve eighty dams, sixty-six hydroelectric stations, and sixty-eight irrigation projects (Hillel 1994). Turkey's goal is to become self-sufficient in food production, and it believes that its water development projects are critical to meeting this goal. While most of the development has been on the Euphrates River, Turkey has plans to develop the Tigris in the year 2000. When the Tigris development is completed, it is expected that one-third of the water supply will be diverted (Hillel 1994). A statement by Turkish prime minister Suleyman Demirel in 1992 clearly indicates Turkey's commitment to the Harmon principle: "We do not say we should share their [Syria and Iraq] oil resources. They cannot say they should share our water resources" (Postel 1993: 16). The potential for conflict is great.

Syria has also developed the Euphrates, although the Syrian projects have not been as successful. The Tabqa Dam was built in 1974 but failed to reach its potential for both irrigation and hydroelectric development. Many of the problems with the dam are the result of policy decisions within Syria. Syria has invested an enormous amount of its gross domestic product (GDP) on security issues, leaving fewer resources available to develop the Euphrates. Further, Syrian strategy includes the overt policy to increase population. Numerical growth is perceived by Syria to be the key to future power and success in the Middle East. Yet, Syria has become a net food importer and has so mismanaged its water supply that the groundwater is also depleted (Hillel 1994). The result of these policies has been that Syria is concerned about any potential loss of water to Turkey. In addition, Syria must be concerned about the consequences of interrupting the flow into Iraq.

By some accounts, Iraq has the most to lose in this transboundary issue. Iraq must absorb losses from both Turkish and Syrian water projects; this is particularly critical since Iraq is already the most arid of the three countries. Further, the water quality is diminished greatly by the time it reaches Iraq, with high concentrations of agricultural chemicals and salts derived from Turkish and Syrian irrigation (Gleick 1994). While Iraq strongly believes it has historical rights to both the Euphrates and Tigris rivers, it recognizes that it is in a poor negotiating position since it is the most downstream state. If Turkey and Syria complete their water projects as planned, Iraq's share of the Euphrates could be reduced as much as 80 percent (Gleick 1994). Iraq is concerned about ensuring its own water security.

There is certainly a great need and potential for cooperation in this case. If the three states follow through with their water projects for both the Tigris and Euphrates, they will exceed the total annual flow of the rivers. Since dams and diversions also increase the rate at which the water evaporates, the states, with their individual water projects, will also reduce the total amount of the resource (Hillel 1994). Yet, there has been limited success in cooperation. Turkey and Syria concluded an agreement in 1987 that guaranteed Syria a set flow from the Euphrates. Syria, in turn, concluded an agreement three years later that guaranteed Iraq a set flow. The cooperation was short lived.

In 1990, Turkey began to divert water from the Euphrates to fill the Ataturk Dam, the key point of the Southeastern Anatolia Project. For one month, Turkey completely shut off the Euphrates, sending no water downstream to Syria and Iraq, although Turkey sent additional water for several months preceding the cutoff. Both Syria and Iraq protested to the Turkish government and threatened military action; Turkey resumed release from the Euphrates. While violence was avoided in this instance, threats of using the water "weapon" to gain political leverage has continued. Turkey threatened to cut off water supply to Syria in 1989 because of its support for the Kurds (Postel 1993). During the crisis in the Persian Gulf, the United Nations considered using Turkish control over the Euphrates as a way to force Iraq to withdraw from Kuwait (Gleick 1994).

The predominance of animosity and conflict in this case has had one rather interesting and unexpected side benefit: the World Bank has stated that it will not support further water projects until the three states reach a mutually acceptable agreement to share the water resources (Hillel 1994). The inability to receive World Bank funding has caused all three states to cease development of their intended water projects. This has not only protected the Tigris-Euphrates system from further alteration, but has also alleviated the immediate pressure for conflict.

▪ Nonrenewable, Boundary Resource: Oil

Technically, oil is a renewable resource; it is created when dead marine microorganisms accumulate on the ocean floor and eventually become released as hydrocarbon molecules. The key word is *eventually;* fossil fuels are called such simply because it takes thousands of years for supplies to become abundant. In fact, some estimate that humans now extract each year what it has taken nature 1 million years to create (Pickering and

Owen 1994). This is the problem: oil reserves are a limited and finite resource, and industrialized states use more of this resource than they can produce themselves. Currently, the United States is the largest user of petroleum, yet the largest oil fields in the United States contain only 4 percent of
the global reserves (Flavin and Lenssen 1991). The United States, as a result, relies on oil imports to fill the gap; most of these imports come from
the Middle East (Flavin and Lenssen 1991). It is estimated that the Middle
East will accommodate 75 percent of U.S. energy needs by the year 2000
(Flavin 1991).

Two issues must be considered when discussing the global oil situation: consumption requirements and domestic availability. Oil consumption is directly tied to development and industrialization. Most major factories and industries rely on oil, as does agribusiness, which relies on
mechanized farm equipment. A state's desire to maintain a strong and
robust military also increases the need for oil, as do citizens' desires for
convenient transportation. These development needs have increased at a
phenomenal rate since World War II. The United States was able to accommodate its needs for many decades; in fact, prior to World War II, the
United States was the primary producer of oil (Võ 1994). However, the
needs soon outpaced the production capability and the U.S. reserves.

More than 50 percent of the global oil reserves are in the Middle East
(Pickering and Owen 1994). Thus, the Middle East has become the site of
an important commercial enterprise that affects both economic development and international politics. The United States has become quite reliant
on Middle Eastern oil; despite its attempt to limit oil imports during the
OPEC crisis in the 1970s, the United States now imports more Middle
Eastern oil than it did before the crisis. Other developed states also depend
on imported oil. Japan has extremely limited reserves of fossil fuels (Pickering and Owen 1994) and must rely almost completely on imports, most
from the Middle East. The Europeans are less reliant on oil than the United
States, but their reserves are also much smaller and more expensive to ex-
ploit (Flavin 1991). However, they still consume more than they produce,
and they are heavily dependent on the Middle East for their supply.

As the situation stands, there are many powerful states in the North
that depend heavily on a resource that is necessary for industrial development, personal consumption, and maintenance of the military. One could
make an argument that the power in these states is determined in large part
by oil. But that resource is very limited and is predominantly found in another region of the world. While this sets up the potential for cooperation,
it has in fact more often led to conflict.

The 1991 Gulf crisis. On August 2, 1990, Iraqi military forces crossed
over the border and seized the sovereign state of Kuwait. While the Iraqi

government made many claims to justify the action, one of the major issues was oil. The Rumailla oil fields are on the border between the two states. Iraq claimed that because Kuwait was using too much of that oil reserve and was stealing oil from the Iraqi side of the border, Iraq should be compensated (Freedman and Karsh 1993). This dispute has its roots in history.

In the nineteenth century, Kuwait was a province of the Kingdom of Iraq. When the Ottoman Empire, which controlled Iraq, ended, Kuwait became a protectorate under Britain and remained so until its independence in 1961. It was later discovered that some of the richest oil reserves were within Kuwaiti territory. Development in Kuwait grew tremendously throughout the 1960s, so that by the 1970s, Kuwait had one of the highest per capita GDPs in the world (Võ 1994). Although the Kuwaiti economy slipped during the 1980s due to inflationary problems, Kuwait was clearly a state that was very well off economically.

While Kuwait was growing at unbelievable rates, Iraq faced its own internal problems. The growth of the Kuwaiti economy was a sore point to Iraq. Iraq still believed that it had historical rights to the territory and threatened to fight Britain during its protectorate period (Freedman and Karsh 1993). The Iraqi economy faced an even greater strain during the 1980s: the war with Iran. The eight-year war left an enormous drain on the Iraqi economy, which was made worse by the fact that Kuwait had supported and lent money to Iran. The economic strain, coupled with disputes over Iraqi rights to Kuwaiti territory, paved the way for the Iraqi invasion in 1990.

The invasion was a clear violation of international law against a sovereign state, and the United Nations responded. Yet, many other clear violations have been ignored by the international community. What made the difference? Many experts suggest that the only reason the United States and the rest of the Northern states took such swift and decisive military action was because of their concern for protecting the oil supply (Võ 1994; Pickering and Owen 1994). The North, especially the United States, had many concerns. It was worried not only that the Kuwaiti oil reserve would be in Iraqi hands, but that Iraq might destroy the oil fields (Warner 1991). Further, there was great concern that Iraq would not stop with Kuwait but would attempt to claim the abundant oil reserves on Saudi territory as well. No matter what the goal of the Iraqis, the North felt that its supply of petroleum was in danger.

What followed was an example of collective security provisions enacted in the United Nations. The UN Security Council condemned the invasion and demanded that Iraq retreat from the borders of Kuwait. When Iraq refused, sanctions were imposed and a multinational military force was created. A short but violent war ensued, with the final result being the Iraqi withdrawal from Kuwait. Although Iraq did destroy some of the oil fields and set as many as 500 oil wells on fire (Warner 1991), the violence produced the desired result: protection of the oil reserves.

One can argue that the actions that took place within the United Nations are an excellent example of cooperation between very different states. However, one can also see that the catalyst for that cooperation was conflict over a limited and very important resource.

■ CONCLUSION: THE NEED FOR SUSTAINABILITY

One of the things that can be learned from the case studies presented in this chapter is that there is the potential for both cooperation and conflict in all natural resource issues. Cooperation is more likely to occur when the parties are otherwise on friendly terms and when the resources are neither severely limited nor critically important. Conflict, particularly violence, is more likely when the parties distrust each other, the resources are both limited and important, and cooperative mechanisms are insufficient.

Perhaps a more important lesson to learn is that there is a great need to alter our use of natural resources to a more sustainable level so that

Source: Ed Stein, *Rocky Mountain News,* 1990. Used by permission of Ed Stein.

cooperation can be encouraged and conflict avoided. The UN World Commission on Environment and Development has defined *sustainable development* as "development which meets the needs of the present without compromising the ability of future generations to meet their own needs" (Elliott 1994). The goal is to ensure that while basic human needs are being met for the present generation, we do not jeopardize the ability of future generations to meet their needs. For the South, sustainable development means finding alternatives to many of the development techniques that are currently depleting the resources. This might result in finding inherent value in standing forests—rather than valuing deforested plots of land—or developing effective fishery management plans so that stocks are not expended. For the North, sustainable development means reducing consumption in general. This would require developing plans for more efficient use of the current resources, as well as exploring more fully solar, wind, and thermal energy sources. Both North and South can cooperate to develop policies that are more efficient and that ensure the provision of basic human needs around the world.

While natural resource issues have been perceived by many policymakers as "low politics," it is clear that natural resources can become "high politics" if the stakes are high enough and the needs critical enough. One solution to avoid conflict and encourage cooperation in natural resource issues is to live more sustainably. This may be in fact the challenge for the twenty-first century.

■ QUESTIONS

1. Use the fourfold matrix (Table 14.1) to determine in which areas cooperation and conflict are most likely to occur.

2. Are there other resources in which the debt-for-nature model might be useful?

3. Should the price of gasoline accurately reflect its cost?

4. If you were president of the United States and faced a threat to Middle Eastern oil reserves similar to the one George Bush faced, what would you do?

5. The United States contains only 5 percent of the global population but uses 25–40 percent of the global resources. Does the United States have a responsibility to try to reduce its national consumption levels? What could it do to accomplish this?

■ SUGGESTED READINGS

Caldwell, Lynton Keith, and Paul S. Weiland (1996) *International Environmental Policy: From the 20th Century to the 21st Century*. Third edition. Durham, NC: Duke University Press.

Earth Works Group (1989) *50 Simple Things You Can Do to Save the Earth*. Berkeley, CA: Earth Works Press.

Elliott, Jennifer A. (1994) *An Introduction to Sustainable Development: The Developing World*. London: Routledge.

Keohane, Robert O., and Marc A. Levy, eds. (1996) *Institutions for Environmental Aid: Pitfalls and Promise*. Cambridge: MIT Press.

Porter, Gareth, and Janet Welsh Brown (1996) *Global Environmental Politics*. Second edition. Boulder, CO: Westview Press.

Switzer, Jacqueline Vaughn (1994) *Environmental Politics: Domestic and Global Dimensions*. New York: St Martin's Press.

Wells, Donald T. (1996) *Environmental Policy: A Global Perspective for the Twenty-first Century*. Upper Saddle River, NJ: Prentice Hall.

Environmental Protection and the Earth Summit: Paving the Path to Sustainable Development

Stephen Collett

While public attention has perhaps focused more on environmental threats to the atmosphere, in the form of global warming and holes in the protective ozone layer, a more immediate menace is the degradation of the earth's land resources. These include soils and related resources of plant and animal life (biodiversity), forests, and freshwater. This is not to say that the climatic and atmospheric problems are not critically serious. However, the impact of the deterioration of land-related resources, if not halted and reversed, will lead sooner to disastrous effects on our growing human population than the three to five decades predicted before climate change may begin. Further, while the intensity and impact of the climate change phenomenon are still speculative, we have graphic and increasingly widespread evidence that the earth's land resources, the very foundation of human society, are being depleted at unsustainable rates.

The environment is at the heart of our visions of peace: green pastures, still waters. It is the means and the mirror by which we know ourselves as individuals and as a species. Solomon's wisdom was marked by his knowledge of the natural life around him: "He discoursed of trees, from the cedar of Lebanon down to the marjoram that grows out of the wall, of beasts and birds, of reptiles and fishes" (1 Kings 4:33 New English Bible). The roots of law in our societies have grown in large part from the need for agreements and standards for sharing and managing land resources. In fact, the Greek stems for our word *economics* refer to the management of the nest.

Still today, the status of a nation's management of its land resources is a good measure of its overall condition and the security of its citizens. Sound environmental management is one of the building blocks of peace, along with respect for human rights, economic and social justice, and democracy. Archaeological evidence suggests that many societies before us—from the Indus Valley (of what is now Pakistan), ancient Rome, and the Yucatan Peninsula—may have disappeared chiefly for ignoring the limitations of their environment. We have come only lately in the modern age to recognize this threat. The Industrial Revolution has seen a period of unprecedented denial of the foundation of our societies in a partnership with nature, or perhaps a reinterpretation that the relationship was chiefly to provide for our unfettered use of the earth's resources. This has been a very serious lapse, given the capacity of our modern global society to cause massive, long-term, and, in some cases, irreversible damage to the environment.

An awareness of the perils we face for ignoring our disruptive impact on natural systems began to take hold only in recent decades. Scientists had been cooperating internationally on environmental matters since the 1950s under United Nations (UN) auspices, but the attention of governments was really engaged only with the 1987 publication of the World Commission on Environment and Development (WCED) report, also known as *Our Common Future*. *Our Common Future* told governments point-blank that the evidence of decades of scientific investigation clearly pointed to a whole web of unsustainable practices whereby our societies are undercutting their own security. The WCED's report defined and illustrated the new concept of *sustainable development* as "development that meets the needs of the present without compromising the ability of future generations to meet their own needs." The discussion it generated in the United Nations led to the UN Conference on Environment and Development, or Earth Summit, in Rio de Janeiro in 1992. Land resources was one of the numerous topics governments examined, and the result was an action plan titled *Agenda 21: Programme of Action for Sustainable Development*.

This chapter examines three major areas that were treated at the Rio summit: (1) protecting biodiversity, (2) combating desertification, and (3) ensuring the sustainable management and protection of forests.

■ BIODIVERSITY

Aside from climate change, the protection of biodiversity was the only issue on the Earth Summit agenda to be negotiated and made ready for adoption at Rio as a binding international legal agreement, or convention. This priority treatment was a reflection of the sharp attention given to biodiversity by *Our Common Future* and the clear realization between both the Southern developing countries and the Northern industrialized countries of the immediate economic and long-term security values of protecting biodiversity. Let us first examine the problems associated with the protection of biodiversity and then look at the solutions offered by the Convention on Biological Diversity (CBD).

Threats to Biodiversity

Biodiversity, the diversity of plant and animal life on earth, is one of the greatest riches of our environment and one we depend on heavily. There are three aspects to the diversity of life-forms: as genes, as species, and as ecosystems. While genetic material contains the key to life-forms, those life-forms must be active themselves in the regeneration and adaptation of the species, and species can exist only in the ecosystems where they are supported in the niches to which they have conformed through time.

Biological diversity is probably being eroded faster today than at any time since the dinosaurs died out some 65 million years ago. Much of what we know about the diversity of life-forms and their loss is estimation, because we lack enough knowledge of both the total numbers of species and microorganisms and their interdependence to know for certain what we are losing and how we might save them. According to estimates from the Global Biodiversity Assessment (UNEP 1996), between 13 and 14 million species live on earth, though only about 13 percent of these have been scientifically described. A good deal more than half the total number of earth's species are found in tropical forests. Both because of the density of species in those ecosystems and because of the rapid changes occurring there, the greatest numbers of species being lost are from the cutting of tropical rain forests. The UN Food and Agriculture Organization (FAO 1991) estimates that 42 million square acres of tropical forests, an area four times the size of Switzerland, are being cleared each year, with probably two-thirds of that for expanding agriculture. The tropical forests give us a dramatic example of the intertwining of fate between human societies, ecosystems, and species.

While the highest number of life-forms lost is from the warm, moist forests, extinction is occurring everywhere. For instance:

- In the decade of the 1980s, at least thirty-four species became extinct in the United States.
- The greatest relative losses recorded recently are on oceanic islands, with some 60 percent of endemic plant species (found nowhere else on earth) in the Galapagos and 75 percent in the Canary Islands endangered (see WCED 1987; WRI 1992; UNEP 1996).

What does this mean for humans? A lot. Virtually all of the main species of ocean fish varieties used for food are estimated to be dangerously overexploited, and freshwater species of fish are endangered in most countries (though the United States has demonstrated an encouraging recent record of recovering health in rivers and lakes).

Loss of genetic diversity could imperil agriculture. The spread of "miracle" varieties of rice, wheat, corn, and other crops since the 1950s, in what has become known as the Green Revolution, has resulted in the loss of traditional local varieties. That is, in the past, each region developed its own seed uniquely fit to its local environment. The Green Revolution imported high-yield varieties that were used over many regions, which resulted in a lack of genetic diversity. With those lost varieties have gone their adapted genetic capacities to withstand disease, drought, flooding, and other threats to the increasing monoculture of today. In Indonesia, 1,500 local rice varieties have become extinct in the past fifteen years. Outbreaks of disease can devastate huge areas where crops are genetically similar, as happened in the Irish potato famine in 1846, the U.S. corn crop in 1970, the Soviet wheat crop in 1972, and citrus crops in Florida in 1984 and Brazil in 1991. In a world with increasing global interdependence and interaction and a decreasing surplus of food stocks and productive land, it is clear that human society is weakening its tolerance for the uncertainties of nature while adding to the disequilibrium.

The food-base reason for protecting biodiversity is the most obvious. But the species with which we share planet earth, from the microscopic life that generates fertility in soils to the mangrove forests that filter the outflow of estuaries, also perform a multitude of important functions that may go unnoticed until it is too late. The depletion of shellfish beds in the Chesapeake Bay means that the waters of the bay, which the shellfish clean through their digestive systems, are circulated at a rate over 100 times slower today than a century ago. The hunting of frogs in India for sending frog legs to European dinner tables has meant mosquitoes in India have lost their natural predators and has brought on a vicious resurgence of malaria when it had almost been eradicated. Plants and animals lost now may have held the key to medicines we will need in the future. The overriding important lesson is that the damage being done today—largely as a result of human activities—will limit the range of options that people will have in the future.

It was with these points in mind, and with the companion realization of a common heritage that will be saved or lost for all of us on earth, that countries moved in 1990 to negotiate a global framework for protecting biodiversity, in preparation for the Earth Summit.

■ A Global Strategy for Protecting Biodiversity

Scientists, diplomats, policymakers, and development experts engaged in drafting an agreement on protecting biodiversity in preparation for the Earth Summit. They set out to develop a legal framework—binding both on national action and in the international arena for those countries adopting the agreement—that would combine and coordinate efforts to save biodiversity in all its forms: as genes, as species, and as habitats. Seven elements of a global strategy were seen as essential (UNEP 1996):

1. Creating conditions and incentives for *local biodiversity conservation*, so that control of the land and its biological resources is in the hands of the people who use them, and so that local knowledge is applied to and local communities benefit from their use
2. Establishing a *national policy framework* for biodiversity conservation, reforming existing public policies that invite waste or misuse of biological resources and promoting conservation and equitable use of biodiversity
3. Creating an *international policy environment* that supports national biodiversity conservation, including a strengthened international legal framework and the application of development assistance to support these ends
4. *Managing biodiversity throughout the human environment* by seeing that biodiversity protection is incorporated not only into public sector but also into private sector planning and development
5. *Strengthening protected areas*, ensuring their sustainability and enhancing their role in biodiversity conservation
6. *Conserving species, populations, and genetic diversity* by strengthening capacities for conservation both in natural habitats and in "banks," like gene banks and zoos
7. *Expanding human capacity* to conserve biodiversity by increasing appreciation and awareness of biodiversity's value and importance, disseminating relevant information, and promoting basic and applied research on biodiversity conservation

After some negotiation, the CBD was signed by 150 heads of government at the Earth Summit and by mid-1997 had been ratified by 165 states. Ratification means that a country's parliament or congress has adopted a convention (or treaty) as its own law of the land and as the framework for international cooperation in the global context. It is noteworthy that while U.S. scientists led much of the work in developing the Global Biodiversity Strategy and though President Clinton has signed the convention, for various political reasons the United States Senate has not yet ratified the CBD. The Clinton administration has said it wants to put this convention before the Senate at the earliest opportunity, and that the United States is undertaking to apply the convention's framework to both U.S. domestic policies and to U.S. foreign assistance as though the convention applied.

■ The Significance of the CBD

The Convention on Biological Diversity is an extremely important model in the effort to put our societies onto a foundation for sustainable development. The idea of sustainable development has taken on the role of a new world view following the Earth Summit. The concept is that humanity must ensure that it meets the needs of its own generation without compromising the ability of future generations to meet their own needs. There are several basic tenets of sustainable development exemplified in the CBD.

Central to global sustainable development is the notion that the states of the industrialized North and those of the developing South must form a *new partnership* in order to reach sustainability together. In the documents of the Earth Summit, the North is recognized as being chiefly responsible for the present state of pollution of the world and for such critical outcomes as climate change and atmospheric ozone depletion. The South, on the other hand, is undergoing an industrial revolution that must not prove as dirty as that of the North. This means that the Northern states must assist their sisters and brothers in the South to achieve a modernization that employs the cleaner new technologies of today and avoids the cheap and dirty path followed in the development of Northern states. To put this another way, if China, Indonesia, India, and Brazil were to cause as much environmental damage on their development path as the Western states did, biodiversity everywhere would suffer. Therefore, a central requirement for the conservation of biodiversity in developing countries is that the issue of poverty there must be addressed and that practical, effective means must be adopted to give those people an alternative to the unsustainable exploitation of their natural resources.

In other words, the South is saying that if you (the North) want us to maintain the gene pools that will someday be of value to you, you are going to have to support us in our overall development strategy, as well as help us manage our biological resources. As to the uses science and industry might find for the biological resources of the South, the convention stipulates that both the knowledge and the benefits of the uses of biological resources must be shared with the countries of origin (where the biological resources originate). In exchange for this agreement to share the rewards of biotechnology, the South—the countries of origin—agree to provide access for scientific investigation under agreed terms.

The CBD was "brought into force" in 1993 after being ratified by the required fifty countries (the number stipulated in the convention). It is being monitored by the Conference of Parties, which is composed of those states that have ratified it. Along with the CBD's importance for its definition of the problems and the recommended approaches to conserving biological diversity, the aspect of North-South partnership written into the Convention on Biological Diversity, and now being practiced under its auspices, is one of the significant achievements of the Earth Summit.

■ DESERTIFICATION

Following the Earth Summit, a second major area for protection of land resources was brought into negotiations for a global convention: combating desertification. Desertification is defined as "land degradation in arid, semiarid and dry sub-humid areas resulting from various factors, including climatic variations and human activities" (UN 1995). That is, desertification is the loss of soils and soil fertility in the drylands of the world. While the real deserts, made of sand or stones and not much else, make up around 16 percent of the drylands, desertification refers to problems over a much wider area, encompassing in all about a quarter of the earth's ice-free land surface. Drylands constitute 66 percent of the continent of Africa, 46 percent of Asia, 75 percent of Australia, and around a third each of Europe and North and South America (Dregne 1991). About a billion people live in these zones and practice rainfed and irrigated agriculture and herding. All three of these farming practices can degrade the soils that support them when they are poorly used.

Desertification has already been intensifying dramatically over the past fifty years. Estimates are that 69 percent of total agricultural lands in dry regions are degraded, which includes 30 percent of irrigated lands, 47 percent of rainfed cropland, and 73 percent of rangelands (Dregne 1991). It should be noted here that the percentage of irrigated lands and rangelands that are degraded in North America is as high as on almost any other continent. Desertification is not only a phenomenon of poor countries, but it is here that the impact is felt most harshly. This is because poor farmers have fewer alternatives to turn to (such as credit, alternative sources of income, or public disaster relief), a narrower margin for survival, and often complicating circumstances such as political unrest or recurrent drought.

Desertification can be viewed as the breakdown of the fragile balance that has allowed plant, animal, and human life to develop together in the drylands. Many factors may contribute to destabilize a society's relationship with the land. These could include prolonged drought, as occurred across the Sahel in Africa in the early 1970s following several decades of excellent rainfall, and the ensuing growth of both human and livestock populations.

A breakdown of the equilibrium between parts of the ecosystem is often self-reinforcing in a downward spiral of destruction for all elements of the life system. Damage to soils occurs when they are made vulnerable to wind and water erosion, when the water table is lowered, or when the ground is choked with salts from poor irrigation practices; it follows that the natural regeneration of vegetation is impaired, and wild animals and livestock compete for remaining forage. Human response is often to intensify usage of the land in an attempt to maintain production levels, converging on dwindling reserves of good land, freshwater, and vegetation. When these strategies fail, families and communities may be left with only the most drastic survival strategy: to sell off or consume their remaining agricultural capital and to migrate, leaving the damaged land behind them.

The greater numbers of those in the outflow from desertified areas are men seeking employment in the informal economic sectors of their nation's cities, or in neighboring countries. Women and children in many of these cases are left to hold their own in the drylands, with fewer to do the labor. While this outcome is surely difficult for all parties, it has been shown that women taking over the management of farming and herding can have a positive effect on the land resources. The women have often been tending to the agriculture anyway, without having the decisionmaking power to make land-use choices as they might see them. Some of the most successful experiences with land recovery and community reorganization in the past decade in the drylands of Africa, northeast Brazil, and India have been with women as farmers, where cooperation among themselves and with those providing outside assistance is a vital factor. This is a very brief and oversimplified version of the causes and outcomes of desertification, but it illustrates the complex mix of issues that need to be addressed in strategies to protect and revive the world's drylands.

■ Combating Desertification

It is possible to reclaim most of the degraded drylands. A good example is the United States' recovery from the "dust bowl" experience of the 1930s, when its midwestern plains were severely damaged by drought and vulnerable to the overextension of grain farming into rainfed dryland. Concerted action to combat desertification has three objectives to be pursued as conditions warrant: (1) to check or prevent desertification of land slightly or not yet degraded; (2) to regenerate, through corrective measures, the productivity of moderately degraded lands; and (3) to restore the productivity of seriously degraded land, using rehabilitation and repair measures.

Accomplishing any of these objectives requires that human beings are seen to be at the heart of the desertification problem, both as perpetrators and as victims. Another way of saying this is that with or without climatic factors, like recurrent drought, human factors clearly underlie many of the physical processes involved in desertification. In cases from around the world, it has been shown that when the causes of desertification are seen in an integrated, comprehensive way and when people are provided with an economic and political environment that supports sustainable management of drylands resources, degraded lands can be restored and communities regenerated. Such factors as land tenure (who owns and makes decisions regarding the use of land), livelihoods for people in dryland communities, the full participation of women, the availability of credit, and the supplies of energy for households, agriculture, and industry are all part of the socioeconomic framework that influences land use. Additionally, there are numerous technologies (both traditional and new) for farming and herding in the drylands that can help to protect and to restore soils and vegetation. In the preparation of the convention, all these aspects were taken together.

As the general plan for combating desertification was being adopted in Rio as a part of Agenda 21, the leaders of African countries made their agreement on other topics, such as support for the conventions on climate change and biodiversity, dependent on a promise from other countries that desertification would be dealt with at a later date in a more comprehensive fashion. African states argued that desertification, affecting more than 100 countries and a sixth of the world's population, should be treated as a global problem—like biodiversity and climate—and that it should be the subject of its own binding international convention. With this agreement, the Earth Summit recommended that negotiations for a Convention to Combat Desertification (CCD) be started as soon as possible after the Rio conference. These talks began in May 1993 and were completed, with a convention in hand, in June 1994.

■ The Convention to Combat Desertification

How could international law have a significant effect on a problem as diffuse (compared with the "pinpoint" sources of emissions of the gases that cause ozone depletion) and as complex as that of desertification? The CCD presents a skilled and creative answer to that question, the result of a wide international effort involving scientists, diplomats, the UN secretariat, and a broad representation of nongovernmental organizations (NGOs) from national and community-based development programs.

The chief components of the convention are as follows:

• The approach should be practical and practicable, with concrete prescriptions on national and regional action programs (that is, within problem areas of Africa, Latin America, Asia, and the Mediterranean region that overlap national boundaries) supported by international cooperative agreements.

• A strong emphasis should be placed on the "bottom-up" approach, meaning that priorities and planning must be based on communities' own perceptions of their problems and preferred solutions. This approach implies that local knowledge and traditions should play their full role, and local participation in decisionmaking should be ensured. The bottom-up approach, while it seems logical in retrospect, is innovative in relation to what were the typical national and international development policies of earlier decades.

• Factors should be emphasized that build the capacities of local communities to manage their resources sustainably, such as providing education and training, reducing poverty, advancing the role of women, and making land tenure rules more equitable.

• Management of resources in drylands should be integrated, whereby land and water management is combined with a concern for energy needs and community incomes; this may include such new elements as the promotion of solar and wind energy, for example.

• A commitment should be made to information exchanges to stimulate a general awareness at all levels—from affected communities to policymakers—of the issues at stake.

• A new partnership should be forged between the poor affected countries of the South and their richer neighbors to the North. From the North this requires a commitment to coordinate better the development assistance, technical support, and market access they can provide. For the affected poor countries, the partnership includes a national responsibility to convene policymaking forums with all partners, from the local to the international levels, and to stand by long-term planning for support and rehabilitation of their drylands.

(Some success stories in combating desertification are presented in Figure 15.1.)

The CCD was warmly received by the member states of the United Nations, both North and South, when it was opened for signature in Paris in October 1994 and signed by 115 countries; by 1997, some sixty-five of those countries had completed their ratification procedures to bring it into force, with the other fifty governments working on that process.

What does this convention give to the three main actors: governments of the North and South and the people at ground level? First, for the countries of the North, the CCD gives a basis for long-term, focused attention to problems in poor countries that have already absorbed a great deal of

---

Figure 15.1   Success Stories in Combating Desertification

The concepts and approaches for dryland management that form the basis for the Convention to Combat Desertification are derived from experiences in addressing the problem in various parts of the world over the past two decades. Some examples follow:

Burkina Faso   Increasing population pressure in the Central Plateau region had, by 1980, produced an area of impoverished soils, declining harvest yields, and scarcities of fuelwood, water, food, and animal fodder: a classic case of progressive desertification.

Building on a regional program of self-help, cooperative planning and resource management, and the clarification of the rights of land-users, by 1993 the following achievements had been gained: soil erosion stabilized on 7,000 hectares of land; grain yields increased from 50 percent to 150 percent; 1,800 kilometers of erosion-control structures built; migration from the region reduced; and the self-esteem of the population and their visions of the future enhanced.

Cape Verde   Between 1985 and 1994, a program initiated jointly by the national governments of Cape Verde and Norway and the UN Sahel Programme addressed the problem of desertification in the S.J. Baptista Valley of this island nation off the west coast of Africa. The valley of some 50 square kilometers has a population of 2,500 on fragile land that was severely eroded due to unsustainable production systems in a climate with low and irregular rainfall.

The implementation of a 1985 master plan for the valley, engaging the local people both as planners and as paid labor, built fifty small irrigation works and one major dam. Not only has the project been successful in renewing areas for cropping and pastoral agriculture, but it has produced substantial income-earning opportunities and established a cooperative framework of future maintenance of the project for the local population.

Australia   The term *Landcare* defines a movement that began in the 1980s in Australia, which now comprises over 1,400 community-based self-help groups for dryland protection and rehabilitation. These groups have been effective in addressing the causes of what had become severe degradation in both pastoral and cropping regions. The Landcare program works to reduce risks associated with change, improve awareness of local problems and joint solutions to be applied, reduce isolation, address off-site effects, and present new technologies as a part of integrated land management packages. The years 1990 to 2000 have been declared the Decade of Landcare, with participation by local land users covering almost 85 percent of the affected areas.

Source: Unpublished country reports to the Intergovernmental Negotiating Committee on a Convention to Combat Desertification (INCD), various years.

---

development aid without making an obvious difference. Continued deterioration of conditions in these regions could lead to problems of civil and political unrest, unsustainable growth in the cities to which dryland residents flee, the loss of potential food-producing lands, and a possibly destabilizing migration across lands and continents, all of which could have wide international impact—in addition to the simple ethic that the human costs of the problem demand attention.

Second, for national governments in the South, the convention can help to initiate a concentrated, long-term effort bringing together all their development partners.

Third, at ground level, on the farms and in the villages of the poor drylands, the convention offers a means to join in a national dialogue and planning process where the perspectives and counsel of those farms and villages are to be given weight. The convention implies a democratization of decisionmaking for national resource use and the offer of training and technology to give local communities the tools with which they can make a difference in their own lives.

The Convention to Combat Desertification is a keystone of the broader effort to achieve sustainable land use and food security for a growing world population. Though there is much hard work ahead for all partners, the CCD gives the legal framework, the long-term commitment, and the vision for promoting a positive change in the drylands.

■ FORESTS

The final example of action stemming from the Earth Summit to protect land resources, and the most recent, is uniting international efforts for the protection and sustainable management of the earth's forests. Like the problem of desertification, forests were given some treatment at the Rio conference, but in an interim manner, realizing that the subject would likely need to be returned to at a later time. The chief problem with negotiations on forests at that time was the suggestion from Northern countries to negotiate agreements on tropical forests alone. As discussed in the previous chapter, the Southern nations did not think that they wanted to lock their resources into legal agreements that would curtail their development options if Northern countries were not similarly bound.

More specifically, while the industrialized countries of the North believe that they are doing fairly well with the protection and sustainable management of their own temperate and boreal forests (though not all environmentalists may agree), they are interested in protecting the tropical forests for two important, and admittedly rather selfish, reasons. First, the great tropical rain forests provide the largest absorption of greenhouse gases of any part of the earth's surface, except perhaps the oceans and seas. Particularly carbon, from the CO_2 released by the burning of fossil fuels in the North, is fixed again in photosynthesis by the tropical forests. Second, as was pointed out early in this chapter, these forests are also the greatest storehouses of biodiversity on earth. Scientists and manufacturers from the North want to be able to investigate these biological resources, in large part to determine their potential uses for commercial exploitation and patenting. While this issue has been worked out in the Convention on Biodiversity in terms of sharing benefits from the uses of such biological resources, the CBD does not address the wider problems of protecting the forests where they are found. In both these cases, tropical forests are providing services—either absorbing greenhouse gases or maintaining a pool of biodiversity—for which people living there are receiving no return.

Meanwhile, the forested developing countries of the South are looking at a vast global trade in forest products (paper, wood, resins, etc.) in which they have only a 20 percent share and the countries of the North enjoy an 80 percent share. Wood and paper, as measured by the market value of companies in that business sector, represent a larger industry than either steel or aerospace in worldwide terms (WBCSD 1996). By comparison, Canada by itself supplies about 10 percent of this trade. For the countries of the South, a global agreement on forests would have to include the means to help them provide their forest dwellers, and those of their people who are cutting the forests to farm, with alternative livelihoods so that old-growth forests (forests that have never been substantially harvested) can be protected. Also, Southern countries would want help to plant plantations of trees to supply their domestic needs for forest products and paper and to allow developing countries to participate with a greater share in world trade from plantation forests. A global agreement tailored to these measurements might satisfy the needs of all parties and at the same time provide a global framework for the protection and sustainable management of all types of forests.

The forests issue contains a differentiation of countries within the North and within the South that adds a twist to this area of negotiations. While all countries experience a rise in consumer demand for wood and paper products that soars as their development curve rises (put another way, paper use is a good indicator of a country's rate of modernization), in both the South and the North there are countries with and without forests. Compare, for example, the Netherlands and Japan (low forest cover) with Canada and Finland (high forest cover). And in the South, compare China or India (low forest cover) with Brazil or Papua New Guinea (high forest cover). This gives us a matrix of four types of countries (see Table 15.1) represented in the forest negotiations: (1) countries with high per capita incomes but little forest; (2) countries with high per capita incomes and large forests; (3) countries with low per capita incomes and low forest cover; and (4) countries with low per capita incomes and high forest cover. With this consideration, the context for forests is not drawn so strictly on a North-South basis as it might at first seem, which makes the politics of the global management of forests more complex.

■ Negotiations on Forests

As of early 1998, negotiations for a global convention on forests had just begun. This is due in some part to the complexities of the issues as described earlier and the difficulties involved in determining what is in a country's own interests. It is also due to the multiple uses forests serve as the densest of earth's ecosystems. For this reason, many of the uses of forests have been covered elsewhere by other agreements, including the

Table 15.1   Country Per Capita Income and Forest Cover

	Low Forest Cover	High Forest Cover
High Income	Netherlands Denmark Germany Japan United Kingdom Environmental Protection[a]	Australia Finland Sweden Canada United States Sustainable development[a]
Low Income	Philippines Somalia Kenya China India Subsistence[a]	Indonesia Brazil Russia Gabon Papua New Guinea Economic development[a]

Source: J. S. Maini, Intergovernmental Forum on Forests, UN Commission on Sustainable Development, New York. Used by permission of J. S. Maini.
Note: a. Areas of priority concern regarding forests.

Convention on Climate Change, the Convention on Biological Diversity, the Convention to Combat Desertification, the International Tropical Timber Agreement, and the Convention on International Trade in Endangered Species. The report of the Intergovernmental Panel on Forests (IPF) that was delivered to the United Nations in 1997 suggests that the nations of the world are being drawn in increasing numbers to the idea that a global convention is needed for forests. This may be a sign of a rising recognition of the fragility and importance of forests (consider, for example, that most of the earth's freshwater resources originate as precipitation on forests) and that a stronger framework is needed to help guide coordination for the sustainable management of forests in their own right.

The IPF meeting between 1995 and 1997 opened several paths toward improving global cooperation for protection and sustainable management of forests. In reporting to the June 1997 "Earth Summit plus Five" meeting (a meeting planned to coincide with the fifth anniversary of the 1992 Earth Summit meeting) of heads of state to review the implementation of the 1992 Earth Summit agreements, the IPF made 135 recommendations for national and international policymaking, cooperation, and controls to manage forests better. The Earth Summit plus Five meeting established a follow-up body, the Intergovernmental Forum on Forests, open to representatives of all concerned governments and charged to work closely with the task force of international agencies assigned to implement the IPF's recommendations and monitor their success. The new Forum on Forests has also been assigned to debate and resolve outstanding issues of trade, financing, and international law affecting the sustainable management of forests and their resources; the forum is to report to the UN Commission on Sustainable Development in the year 2000.

■ LEGAL INSTRUMENTS FOR SUSTAINABLE MANAGEMENT OF LAND RESOURCES

In closing this brief review of the formation and use of international legal instruments for environmental protection and sustainable development, we might ask: What does a *convention* contribute to the pressing and difficult challenges of sustainability? In the three examples of international agreements discussed in this chapter (and those in other chapters concerning human rights, peacemaking, air pollution, arms control, etc.), we have seen examples of the following functions that the negotiation and implementation of global legal instruments can serve:

• *Building consensus.* The process of drafting international legal instruments calls into play an extensive process wherein political, social and economic, scientific, and legal issues are investigated, reviewed, debated, and finally negotiated. Within the environment/development sphere, every aspect since the beginning of preparations for the Earth Summit has been produced by a consensus process; that is, the issues are discussed and decisions debated until all parties are on board with their concerns recognized. Both in preparation for and in follow-up to the Earth Summit, consensus building has proven to be a rich tool for increasing understanding and cooperation. The products are a new range of North-South partnerships and an understanding of common but differentiated responsibilities in the global effort.

• *National adoption.* The process of gaining national ratification and implementation for international legal instruments has served as a useful guide for policy planning at the national level.

• *Creation of a permanent forum.* The "conference of parties" formed of those states that have ratified a convention serves as a permanent forum for monitoring and supporting implementation of its results and makes the adjustments and additions that they may find are necessary to that end.

• *Complementary functions.* Under these conventions, it is usual to create subsidiary bodies to provide services necessary to the conference of parties. These include scientific and technical advisory bodies to the convention, financial mechanisms to generate and channel resources for implementation of the convention, and institutional support for exchanges of technology between country parties to the convention.

• *Awareness raising.* An important function served by the international legal instruments for the environment, in addition to their central purpose of establishing standards and systems for cooperation and policy coordination between nations, is that they draw attention to the primary issues, both from policymakers at local, national, and international levels and from the general public. Education and awareness raising are obviously of crucial importance if the overall task is to change the way our societies perceive their role in relation to natural resources and systems.

Immense progress has been made for raising awareness and providing legal frameworks for sustainable environmental management in the past decade. There are a number of areas, such as forests and freshwater, where the job has only been begun. Where the framework for protection and sustainable use exists, such as with biodiversity and drylands, the continuing challenge is to achieve both international support and local implementation. The task ahead lies in seeing that these international agreements are widely adopted and integrated into national policies and programs and that they lead to real innovative partnerships within and among nations—north, south, east, and west.

■ QUESTIONS

1. Why are biodiversity, desertification, and destruction of rain forests perceived as global issues?

2. Why can giving women more decisionmaking power have a positive impact on the environment in various situations?

3. Why are tropical forests of interest both to those countries where they are found and to the industrialized states of the North?

4. How is it in the interest of countries to sign international environmental agreements?

■ SUGGESTED READINGS

Bender, William H. (1997) "How Much Food Will We Need in the 21st Century?" *Environment* 39, no. 2.

Collett, J., and S. Karakasian (1996) *Greening the College Curriculum: A Guide to Environmental Teaching in the Liberal Arts*. Covelo, CA: Island Press.

Dommen, Edward (1993) *Fair Principles for Sustainable Development: Essays on Environmental Policy and Developing Countries*. Brookfield, VT: Ashgate.

FAO Position Paper (1993) "Sustainable Development of Drylands and Combatting Desertification." Rome: Food and Agriculture Organization.

United Nations (1992) *Agenda 21, The United Nations Programme of Action from Rio*. New York: United Nations.

———— (1992) *Convention on Biological Diversity*. New York: United Nations.

———— (1995) *Convention to Combat Desertification and Drought, Particularly in Africa*. New York: United Nations.

UN Environment Programme (1994) *Our Planet: Special Issue on Desertification* 6, no. 5.

World Commission on Environment and Development (1987) *Our Common Future*. New York: Oxford University Press.

World Conservation Union (1991) *Caring for the Earth: A Strategy for Sustainable Living*. Gland, Switzerland: World Conservation Union.

World Resources Institute (1992) *Global Biodiversity Strategy*. Washington, DC: World Resources Institute.

Part 5

Conclusion

Future Prospects

Michael T. Snarr

One question of interest to people studying global issues is, given the critical nature of these issues, what will the future look like? Will humans devise methods for dealing more effectively with global issues? Will conditions get better or worse? Four of the more popular scenarios of what the world might look like in the year 2030 or so describe world government, regionalism, decentralization, and the status quo.

■ WORLD GOVERNMENT

Some scholars argue that a world government, consisting of a powerful central actor with a significant amount of authority, is the method by which we will organize ourselves in the future. The increasing degree of consensus being reached on economic issues, including international agreements like the General Agreement on Tariffs and Trade (GATT) and the North American Free Trade Agreement (NAFTA), is often cited as evidence that a world government is in fact a possibility. Similarly, for those arguing that the world is moving toward a single global culture (i.e., McWorld), a world government might not seem beyond reach.

In contrast to a slow, evolutionary movement toward consensus on issues like economics, it is possible that a world government might be created after a catastrophe. An exchange of nuclear weapons or a dramatic increase in ozone depletion might shock governments into calling for a central authority that would avoid long, drawn-out negotiations among 180 or more sovereign states.

An obvious problem with the world-government scenario is the un-likeliness that the countries of the world would voluntarily give up their sovereignty. Furthermore, a world government would face many practical problems, such as who would be responsible for enforcing laws. Would a world government have a powerful military? If so, the fear of tyranny would be realistic. If not, its enforcement capabilities would be questionable.

There are other possibilities in addition to a true world government. A *federation* would establish a relatively weaker world government, simi-lar to the model of the United States, where the federal government shares power with the states. Even weaker would be a *confederation*, in which states would be the dominant actors but would give the world government some jurisdiction. Both federate and confederate systems would give a world government more power than the United Nations currently possesses.

■ REGIONALISM

In the regionalism scenario, countries are organized into groupings based on geographic proximity, perhaps following the pattern of current economic group-ings like NAFTA, the European Union (EU), and the Asia Pacific Economic Cooperation (APEC) forum. As with a world government, countries would not completely relinquish their sovereignty, but that sovereignty would likely be sig-nificantly reduced. The European Union is the leader in the movement toward economic and political cooperation. Not only has the EU drastically reduced barriers to trade and to the movement of people within its borders, it has also made progress toward a single economic currency and a common foreign poli-cy. Although NAFTA and APEC are relatively young in comparison to the EU, their formation represents the current popularity of regional arrangements.

Of course, the regionalism scenario also must deal with the reluctance of countries to relinquish their sovereignty, the fear of concentrating too much power in the hands of a central government, and so on. However, these issues may be easier to resolve in smaller groupings of states than in a world-government context.

On the positive side, regionalism would facilitate the coordination of regional policymaking on global issues such as the environment, human rights, and trade. Still, the enhanced ability of countries to coordinate poli-cies within their respective regions would not necessarily translate into co-operation between regions. It has been argued that regionalism would sim-ply transform a world in which *countries* compete into one in which *regions* compete, without solving pressing global problems.

■ DECENTRALIZATION

At the same time that free trade and environmental agreements are being enacted, there is significant evidence of decentralization or disintegration. One example is the strong separatist movement mounted by Canada's French-speaking province of Quebec. Another is the incident in Texas, in 1997, when an armed group of individuals took hostages and demanded that Texas become a separate country. Although the various separatist movements have differing motives, as we saw in Chapter 3, many of them do have in common a desire for self-determination—that is, the desire to break away from the dominant culture and govern themselves. If many of these groups succeed, instead of "one world" or a few regions, hundreds (or perhaps thousands) of new countries could emerge. Each new country would, of course, be smaller and more culturally homogeneous than today's countries, which ideally would alleviate some of the tensions discussed in Chapter 3. However, it also would make achieving international consensus on issues like the environment, human rights, and nuclear proliferation more difficult.

These sorts of disintegrative movements have a negative connotation since, in many cases, they involve violence. There is, however, another type of locally oriented movement, commonly referred to as *civil society,* that has gained momentum in recent years. Civil society comprises nongovernmental, nonprofit groups such as social service providers, foundations, neighborhood watch groups, and religion-based organizations. India's Chipko movement, Kenya's Green Belt movement, the Grameen Bank (all discussed in Chapter 10), and Habitat for Humanity are examples of civil society, or *grassroots movements.* In recent decades, more and more people have turned to civil society, rather than government, to solve their problems.

Reading this book, you may have noticed the many global nongovernmental organizations (NGOs) mentioned. The number of NGOs has increased dramatically, from about 200 in the early 1900s to nearly 5,000 at the end of the century. Their ranks include Amnesty International, Greenpeace, CARE, and the World Wildlife Federation. Composed of private citizens in more than one country, they focus on such global issues as the environment, poverty, human rights, and peace.

Those frustrated with government's inability to solve global problems insist that centralized governments are not the most effective way to deal with these problems. Governments, they argue, are simply too far removed from local communities to understand completely the nature of a particular problem and to offer effective solutions. Advocates of civil society are encouraged by the dramatic increase in NGOs. Critics, however, believe local, grassroots efforts will be insufficient to solve global problems like nuclear proliferation, ozone depletion, and global warming. They argue that governments are the only actors with sufficient resources to effectively confront these large-scale issues.

■ STATUS QUO

Perhaps the most likely scenario is one in which no dramatic changes occur over the next several decades. This is not to say that change will be absent, but that it will be only a gradual continuation of current trends toward globalization in the areas of economic integration (GATT and NAFTA), information flow between countries, the importance of nongovernmental actors (including multinational corporations), and cooperation among countries on environmental and other issues. Citizens will continue to pledge their allegiance to countries, not economic blocs; states, not groups of private citizens, will remain the dominant political actors; and short-term domestic interests will prevent states from surrendering their sovereignty.

Does this scenario allow an effective response on the part of the global community to the issues discussed in this book? Critics view the status quo with suspicion because it has made disappointing progress thus far in such areas as global warming, peacemaking, and poverty, especially poverty among children.

■ THE FUTURE: SOURCES OF HOPE AND CONCERN

As many of the chapters in this book point out, there are positive signs in the world's attempts to deal with the multitude of pressing global issues. Smallpox has been eliminated. There is a cooperative effort to deal with ozone depletion. Women have been increasingly successful in forming effective grassroots movements. In the developing world, the United Nations Development Programme (UNDP 1996: 20–21) reports that in the last two to three decades:

- the lives of 3 million children per year have been saved by immunization programs
- the infant mortality rate has been cut by more than one-half
- school enrollment at the primary level has increased from 48 percent to 77 percent
- access to safe water has increased from 36 percent to almost 70 percent

- life expectancy has increased by approximately 35 percent; in many countries it is now more than seventy years

In the industrialized countries:

- the number of women in the workforce is over 40 percent
- inflation has been kept under control
- energy use has been significantly reduced
- in nearly every country the life expectancy exceeds seventy-five years

Despite these successes, there is still a long way to go on a number of issues—many challenges still exist to the creation of a better world. In the remainder of this chapter, two such challenges are discussed.

Perhaps one of the biggest obstacles we face is created by the South's desire for economic growth coupled with the North's overconsumption. As we have seen, the North, which constitutes only 15 percent of the world's population, is responsible for creating the vast majority of our environmental problems (acid rain, ozone depletion, greenhouse gases, resource depletion, etc.). Meanwhile, the South, which comprises 85 percent of the world's population, is trying to emulate the North. If the South industrializes in the same fashion as the North, what will the environmental consequences be? The prospect of hundreds of millions of Chinese driving polluting automobiles rather than riding bicycles is a prospect that concerns many.

At the heart of this issue is *sustainable development*—the idea that development today should not negatively affect the lives of future generations. Environmentalists have stressed that long-term interests must be given higher priority; for example, forests should not be clear-cut in order to obtain short-term profit, since the indiscriminate clearing of forests is environmentally harmful and will inhibit future development. The idea of sustainable development has been largely accepted in the North, but has received criticism from the countries of the South. The latter argue that the North, which is responsible for the vast majority of the world's environmental problems, has no moral right to tell the South that it cannot follow the North's development path. These same critics argue that in many poor countries survival is often at stake, so that environmental concerns must be tied to the issue of current development. Based on this premise, the South insisted that the 1992 Earth Summit focus not only on the environment but also on development, and as a result, the conference was officially called the United Nations Conference on Environment and Development.

One possible approach to this complex issue is the transfer of advanced, "environmentally friendly" technology from the North to the

South. This might allow the South to avoid the adverse affects that accompanied Northern development. The South, however, is not in a position to purchase this technology, and the North has balked at significant transfers. Currently, there is no easy solution to this contentious problem.

A related issue is widespread poverty, another enormous obstacle to a better world. Several chapters in this book highlighted the connection between poverty and other issues. Chapter 12 pointed out that health is directly related to poverty—that the poorer you are, the more likely you are to suffer from disease or malnutrition. Chapters 9 and 10 discussed how the number of children a woman bears will decrease as poverty is alleviated and women gain more control over their lives. We were reminded in Chapters 13–15 that for those who are desperately poor, issues of immediate survival must take precedence over concerns about the environment. Chapter 2 underscored the vast amounts of money spent on military budgets at the expense of social programs such as health and education.

Central to the issue of poverty is the unequal distribution of wealth, which appears to be getting worse. Chapter 7 explained that the net flow of money is going from South to North (also see Figure 8.2). At the same time, as those who live in the North know, poverty is not simply a question of North-South relations: there are many pockets of poverty within the wealthy countries, and evidence suggests that the gap between the rich and poor is increasing within countries as well as between them.

Finding a solution to this problem will be difficult. At the domestic level, as Chapter 8 pointed out, a country must have economic growth before income can be redistributed; but economic growth does not guarantee better income distribution. Chapter 11 demonstrated that focusing on taxation, education, health care, and other such issues is necessary to foster a more favorable distribution of wealth; however such an approach typically has little support among those whose wealth would be transferred. The issue becomes even more complex if we confront the *global* distribution of wealth. Within a single country, the wealthy are often taxed at higher rates to support social programs. To attempt to tax the wealthy countries in order to pay for social programs in poorer countries would meet a great deal of opposition, not only from the wealthy in the North, but also from the middle- and lower-income populations. Historically, voluntary aid from North to South has helped somewhat, but has been insufficient to seriously address the poverty gap; also, as suggested in Chapter 7, traditional aid may not be the most effective approach to fostering development.

It is important to recognize that the future has not yet been written. The choices that governments, NGOs, and individuals make will have a critical effect on the issues discussed here. Assuming you agree that these issues deserve serious attention, whether on the grounds of religion, humanitarianism, patriotism, or self-interest, the practical question remains: "What can I do to make a positive difference?" A few suggestions follow.

Addressing Political Issues

- write to your government representatives—this can be more effective than you think
- form, join, or support an NGO—important changes have come about as the result of grassroots movements
- vote

Addressing Environmental Issues

- ride a bicycle or walk to work and school—this can reduce the amount of pollution put into the atmosphere, save you money, and lead to better health
- buy local produce—shipping food from across the country or from another country contributes to pollution
- buy recycled products or those with less packaging
- wear an extra layer of clothing and turn down the thermostat in the winter, turn the lights off when you're not using them, and turn off the water while you are brushing your teeth
- vote

This list is far from exhaustive, and you can find many good books written on the subject (see, for example, Earth Works Group 1989).

Although national and local (and perhaps regional) governments will continue to play important roles, we cannot depend on them to solve all of the problems discussed in this book. It is up to each individual to work to create the world he or she prefers. "The most revealing world order statement each of us makes is with his or her life" (Falk, Kim, and Mendlovitz 1982: 14). However, dramatic results at the global level will be seen only when positive change is adopted and promoted by many of the world's citizens—a strong motivation for those with knowledge of these issues to become active and educate others.

■ QUESTIONS

1. Which of the future world orders do you think is most likely to emerge? Which do you think is most desirable?

2. Can you think of another possible world order?

3.   What serious challenges, in addition to poverty and the need for sustainable development, do you think confront humanity?

4.   What items would you add to the list of things you can do as an individual to make the world a more livable place?

■ SUGGESTED READINGS

Brown, Lester R., ed. *State of the World* (annual). New York: W.W. Norton.
Commission on Global Governance (1995) *Our Global Neighborhood.* New York: Oxford University Press.
Earth Works Group (1989) *50 Simple Things You Can Do to Save the Earth.* Berkeley, CA: Earth Works Press.
Pirages, Dennis C. (1996) *Building Sustainable Societies: A Blueprint for a Post-Industrial World.* Armonk, NY: M.E. Sharpe.
Simai, Mihaly (1994) *The Future of Global Governance: Managing Risk and Change in the International System.* Washington, DC: United States Institute of Peace Press.
United Nations (1990) *Global Outlook 2000.* New York: United Nations.

Bibliography

Ahmed, Samir (1994) "Principles and Precedents in International Law Governing the Sharing of Nile Water." In P. P. Howell and J. A. Allan, eds. *The Nile: Sharing a Scarce Resource*. New York: Cambridge University Press.

Altman, D. G., et al. (1996) "Tobacco Promotion and Susceptibility to Tobacco Use Among Adolescents," *American Journal of Public Health* 86, no. 11.

Amler, R. W., and H. B. Dull (1987) *Closing the Gap*. Oxford: Oxford University Press.

Anand, Anita (1983) "Saving Trees, Saving Lives: Third World Women and the Issue of Survival." In Leonia Caldecott and Stephanie Leland, eds. *Reclaim the Earth: Women Speak Out for Life on Earth*. London: Women's Press.

Aspin, Les (1994) *Secretary of Defense Annual Report to the Congress and President*. Washington, DC: Government Printing Office.

Barber, Benjamin R. (1992) "Jihad vs. McWorld," *Atlantic Monthly*, March.

——— (1996) *Jihad vs. McWorld*. New York: Ballantine Books.

Bates, A. K. (1990) *Climate in Crisis*. Summertown, TN: Book Publishing Company.

Bell, Daniel A. (1996) "The East Asian Challenge to Human Rights: Reflections on an East West Dialogue," *Human Rights Quarterly* 18, no. 3.

BFW (Bread for the World Institute) (1994) *Hunger 1995: Causes of Hunger*. Washington, DC: BFW Institute.

——— (1995) "At the Crossroads: The Future of Foreign Aid." Occasional Paper, No. 4. Washington, DC: Bread for the World Institute.

——— (1997) *Hunger in a Global Economy: Hunger 1998*. Washington, DC: BFW Institute.

Birdsall, Nancy, Thomas Pinckney, and Richard Sabot (1996) "Why Low Inequality Spurs Growth: Savings and Investment by the Poor." Inter-American Development Bank Working Paper Series, No. 327. Washington, DC: IDB.

Birdsall, Nancy, David Ross, and Richard Sabot (1995) "Inequality and Growth Reconsidered: Lessons from East Asia," *World Bank Economic Review* 9, no. 3.

Bojtar, Endre (1988) "Eastern or Central Europe?" *Cross Currents* (a yearbook of Central European culture) 7.

Boserup, Ester (1970) *Women's Role in Economic Development*. New York: St. Martin's Press.

———— (1981) *Population and Technological Change: A Study of Long Term Trends*. Chicago: University of Chicago Press.

Boston Globe (1996) Special Report, "Armed for Profit: The Selling of U.S. Weapons," February 11.

Bouhdiba, Abdelwahab (1982) *Exploitation of Child Labour*. New York: United Nations.

Boulding, Elise (1992) *The Underside of History: A View of Women Through Time*. Revised edition. Newbury Park, CA: Sage Publications.

Boutros-Ghali, Boutros (1992) *An Agenda for Peace: Preventive Diplomacy, Peacemaking and Peace-keeping*. New York: United Nations.

———— (1995) *An Agenda for Peace*. Second edition. New York: United Nations.

Braudel, F. (1981) *The Structures of Everyday Life: Civilization and Capitalism— 15th–18th Century*. Vol. 1. New York: Harper & Row.

Breuilly, John (1993) *Nationalism and the State*. Second edition. Chicago: University of Chicago Press.

Brown, Lester R., ed. (1993) "A New Era Unfolds." In Lester R. Brown, ed. *State of the World*. New York: W. W. Norton.

————, ed. (1996) "The Acceleration of History." In Lester R. Brown, ed. *State of the World*. New York: W.W. Norton.

Bundy, McGeorge, William J. Crowe, Jr., and Sidney Drell (1993) *Reducing Nuclear Danger: The Road Away from the Brink*. New York: Council on Foreign Relations Press.

Burstyn, Linda (1995) "Female Circumcision Comes to America," *Atlantic Monthly*, October.

Byrd, Veronica (1994) "The Avon Lady of the Amazon," *Business Week*, October 24.

Caldwell, John C. (1982) *Theory of Fertility Decline*. New York: Academic Press.

Carson, Rachel (1962) *Silent Spring*. New York: Fawcett Crest.

Castles, Stephen, and Mark J. Miller (1993) *The Age of Migration: International Population in the Modern World*. New York: Guilford Press.

Cavanaugh, John, et al., eds. (1992) *Trading Freedom: How Free Trade Affects Our Lives, Work, and Environment*. San Francisco: Institute for Food and Development Policy.

CGG (Commission on Global Governance) (1995) *Our Global Neighborhood*. New York: Oxford University Press.

Claude, Richard Pierre, and Burns H. Weston, eds. (1992) *Human Rights in the World Community*. Philadelphia: University of Pennsylvania Press.

Columbus Dispatch (1993) "U.N. Group Begins War on Female Circumcision," May 13.

Commoner, Barry (1992) *Making Peace with the Planet*. New York: New Press.

Crossette, Barbara (1996a) "U.N. Endorses a Treaty to Halt All Nuclear Testing," *New York Times*, September 11.

———— (1996b) "World Is Less Crowded Than Expected, the U.N. Reports," *New York Times*, November 17.

Davis, Zachary S. (1991) *Non-Proliferation Regimes: A Comparative Analysis of Policies to Control the Spread of Nuclear, Chemical, and Biological Weapons and Missiles*. Washington, DC: Congressional Research Service.

DeMont, John (1995a) "Gunboat Diplomacy," *Maclean's*, March 20, 1995.

———— (1995b) "Reeling in a Deal," *Maclean's*, April 24, 1995.

Desai, Narayan (1972) *Toward a Nonviolent Revolution*. Varanasi, India: Sarva Seva Sangh Prakashan.

Dodge, Robert (1994) "Grappling with GATT," *Dallas Morning News*, August 8.

Donnelly, Jack (1993) *International Human Rights*. Boulder, CO: Westview Press.

Dregne, H. E., ed. (1991) *Degradation and Restoration of Arid Lands*. Lubbock: Texas Tech University, International Center for Arid and Semiarid Land Studies.

Drêze, Jean, and Amartya Sen (1989) *Hunger and Public Action*. Oxford: Clarendon Press.

Drinan, Robert F. (1987) *Cry of the Oppressed*. San Francisco: Harper & Row.

Earth Works Group (1989) *50 Simple Things You Can Do to Save the Earth*. Berkeley, CA: Earth Works Press.

Ehrlich, Paul, and Anne Ehrlich (1992) *The Population Explosion*. New York: Doubleday.

Elliott, Jennifer A. (1994) *An Introduction to Sustainable Development: The Developing World*. London: Routledge.

Faison, Seath (1997) "China Turns the Tables, Faulting U.S. on Rights," *New York Times*, March 5.

Falk, Richard, Samuel S. Kim, and Saul H. Mendlovitz (1982) *Toward a Just World Order*. Vol. 1. Boulder, CO: Westview Press.

Fallows, James (1993) "How the World Works," *Atlantic Monthly*, December.

FAO (Food and Agriculture Organization) (1991) *Second Interim Report on the State of the World's Forests*. Rome: FAO.

——— (1996) *The Sixth World Food Survey*. Rome: FAO.

Farer, Tom J. (1992) "The United Nations and Human Rights: More Than a Whimper, Less Than a Roar." In Richard Pierre Claude and Burns H. Weston, eds. *Human Rights in the World Community*. Philadelphia: University of Pennsylvania Press.

Felice, William F. (1996) *Taking Suffering Seriously*. Albany: State University of New York Press.

Fiske, Edward B. (1993) *Basic Education: Building Block for Global Development*. Washington, DC: Academy for Educational Development.

Flavin, Christopher (1991) "Conquering U.S. Oil Dependence," *Worldwatch* 4, no. 1.

——— (1996) "Facing Up to the Risks of Climate Change." In Lester R. Brown, ed. *State of the World*. New York: W.W. Norton.

Flavin, Christopher, and Nicholas Lenssen (1991) "Designing a Sustainable Energy System." In Lester R. Brown, ed. *State of the World*. New York: W.W. Norton.

Forsberg, Randall, William Driscoll, Gregory Webb, and Jonathan Dean (1995) *Nonproliferation Primer: Preventing the Spread of Nuclear, Chemical, and Biological Weapons*. Cambridge: MIT Press.

Forsyth, Randall W. (1996) "The End of Communism, the End of History and Now, the End of Business Cycles?" *Barron's*, November 18.

Forsythe, David P. (1991) *The Internationalization of Human Rights*. Lexington, MA: Lexington Books.

Freedman, Lawrence, and Efraim Karsh (1993) *The Gulf Conflict, 1990–1991: Diplomacy and War in the New World Order*. Princeton: Princeton University Press.

Friedman, Thomas L. (1997) "Rethinking China, Part I." *New York Times*, March 3.

Galtung, Johan (1969) "Violence, Peace, and Peace Research," *Journal of Peace Research* 6, no. 6.

Garrett, Laurie (1994) *The Coming Plague.* New York: Farrar, Straus & Giroux.

Ghazi, Polly, Frank Smith, and Claire Trevena (1995) "Our Plundered Seas," *World Press Review* 42, no. 6.

Gleick, Peter H. (1994) "Water, War and Peace in the Middle East," *Environment* 36, no. 3.

Gore, Al (1992) *Earth in the Balance: Ecology and the Human Spirit.* Boston: Houghton Mifflin.

Graham, Edward M. (1996) *Global Corporations and National Governments.* Washington, DC: Institute for International Economics.

Gribbin, J. (1988) *The Hole in the Sky: Man's Threat to the Ozone Layer.* New York: Bantam Books.

Harbottle, Michael (1971) *The Blue Berets.* London: Leo Cooper.

Harper, Charles L. (1995) *Environment and Society: Human Perspectives on Environmental Issues.* Upper Saddle River, NJ: Prentice Hall.

Hartung, William D. (1995) *And Weapons for All.* New York: HarperCollins.

Hauchler, Ingomar, and Paul M. Kennedy, eds. (1994) *Global Trends: The World Almanac of Development and Peace.* New York: Continuum.

Hersh, Seymour M. (1994) "The Wild East," *Atlantic Monthly,* July.

Higgins, Rosalyn (1996) *United Nations Peacekeeping 1946–1967.* Vol. 1, *The Middle East.* London: Oxford University Press.

Hillel, Daniel (1994) *Rivers of Eden: The Struggle for Water and the Quest for Peace in the Middle East.* New York: Oxford University Press.

Holdren, John P., and Paul R. Ehrlich (1974) "Human Population and the Global Environment," *American Scientist* 62 (May).

Holm, Hans-Henrik, and Georg Sørensen, eds. (1995) *Whose World Order? Uneven Globalization and the End of the Cold War.* Boulder, CO: Westview Press.

Human Rights Watch (1988) "The Persecution of Human Rights Monitors: December 1987 to December 1988, A Worldwide Survey." New York: Human Rights Watch.

Huntington, Samuel P. (1996) "The West: Unique, Not Universal," *Foreign Affairs* 75, no. 6.

IATA (International Air Transport Association) *World Air Transport Statistics: Annual Reports.* Geneva: IATA.

ICRC (International Committee of the Red Cross) (1993) "A Time for Decision," *International Review of the Red Cross* (November-December).

ILO (International Labour Organization) (1976) *Wages and Working Conditions in Multinational Enterprises.* Geneva: ILO.

——— (1993) *World Labour Report.* Geneva: ILO.

——— (1996) *World Employment: 1996/97.* Geneva: ILO.

IMF (International Monetary Fund) (1991) *International Capital Markets, Developments and Prospects.* Washington, DC: IMF.

Independent Commission on Disarmament and Security Issues (1982) *Common Security.* New York: Simon & Schuster.

Independent Commission on International Development Issues (1980) *North-South: A Programme for Survival.* Cambridge: MIT Press.

IPA (International Peace Academy) (1984) *Peacekeeper's Handbook.* New York: Pergamon Press.

Iyer, Pico (1993) "The Global Village Finally Arrives," *Time* 21, no. 142 (special issue).

Jenkins, Christopher, et al. (1997) "Tobacco Use in Vietnam," *Journal of the American Medical Association* 277, no. 21.

Jones, Jeffrey R. (1992) "Environmental Issues and Policies in Costa Rica: Control of Deforestation," *Policy Studies Journal* 20, no. 4.

Karp, Aaron (1994) "The Arms Trade Revolution: The Major Impact of Small Arms," *Washington Quarterly* 17 (Autumn).

Kegley, Charles W., and Eugene R. Wittkopf (1997) *World Politics: Trend and Transformation.* New York: St. Martin's Press.

Kent, George (1995) *Children in the International Political Economy.* New York: St. Martin's Press.

Kerr, R. A. (1989) "Greenhouse Skeptic Out in the Cold," *Science,* December.

Klinger, Janeen (1994) "Debt-for-Nature Swaps and the Limits to International Co-operation on Behalf of the Environment," *Environmental Politics* (Summer).

Kohn, Hans (1965) *Nationalism: Its Meaning and History.* Revised edition. New York: Van Nostrand Reinhold.

Korten, David C. (1996) *When Corporations Rule the World.* West Hartford, CT: Kumarian Press.

Lamar, B. (1991) "Life Under the Ozone Hole: In Chile, the Mystery of the Bug-eyed Bunnies," *Newsweek,* December 9.

Landres, Shawn (1996) "For God and Country: The Importance of Religion in the Study of Nationallities," *ASNews: The Newsletter of the Association for the Study of Nationalities* 2, no. 3 (Fall).

Laurance, Edward J. (1992) *The International Trade in Arms,* New York: Lexington Books.

Leggett, J. (1990) "The Nature of the Greenhouse Threat." In J. Leggett, ed. *Global Warming: The Greenpeace Report.* New York: Oxford University Press.

Liebich, André, Daniel Warner, and Jasna Dragovic, eds. (1995) *Citizenship East and West.* London: Kegan Paul International.

Lindzen, R. (1993) "Absence of Scientific Basis," *Research and Exploration* (Spring).

Lutz, Wolfgang (1994) "World Population Trends: Global and Regional Interactions Between Population and Environment." In Arizpe M. Lourdes, Priscilla Stone, and David C. Major, eds. *Population and Environment: Rethinking the Debate.* Boulder, CO: Westview Press.

Mahony, Rhona (1992) "Debt-for-Nature Swaps: Who Really Benefits?" *Ecologist* 22, no. 3.

Malthus, T. R. (1878) *An Essay on the Principle of Population.* Eighth edition. London: Reeves & Turner.

McCarthy, Sheryl (1996) "Fleeing Mutilation, Fighting for Asylum," *Ms.,* July-August.

McGwire, Michael (1994) "Is There a Future for Nuclear Weapons?" *International Affairs* 70, no. 2.

McKibben, B. (1989) *The End of Nature.* New York: Anchor Books.

McKinney, M. L., and R. M. Schoch (1996) *Environmental Science: Systems and Solutions.* Minneapolis/St.Paul: West Publishing.

McNaugher, Thomas L. (1990) "Ballistic Missiles and Chemical Weapons," *International Security* 15, no. 2.

Meadows, D. H., D. L. Meadows, and J. Rander (1992) *Beyond the Limits: Confronting Global Collapse, Envisioning a Sustainable Future.* Mills, VT: Chelsea Green Publishers.

Michaels, P. (1992) *Sound and Fury: Science and Politics of Global Warming.* Washington, DC: Cato Institute.

Moon, Bruce E. (1996) *Dilemmas of International Trade.* Boulder, CO: Westview Press.
———— (1998) "Exports, Outward-Oriented Development, and Economic Growth," *Political Research Quarterly* (March).
Mowlana, Hamid (1995) "The Communications Paradox," *Bulletin of Atomic Scientists* 51, no. 4.
Murray C. J. L., and A. D Lopez, eds. (1996) *Global Burden of Disease.* Cambridge: Harvard University Press.
Narayan, Desai (1972) *Toward a Nonviolent Revolution.* Varanasi, India: Sarva Seva Sangh Prakashan.
National Research Council (1986) *Population Growth and Economic Development: Policy Questions.* Washington, DC: National Academy Press.
New York Times (1996) "Third World Debt Crisis," June 28.
Nincic, Miroslav (1982) *The Arms Race: The Political Economy of Military Growth.* New York: Praeger.
Payne, Stanley (1995) *A History of Fascism 1914–1945.* Madison: University of Wisconsin.
Pickering, Kevin T., and Lewis A. Owen (1994) *An Introduction to Global Environmental Issues.* London: Routledge.
Polanyi, Karl (1944) *The Great Transformation.* New York: Farrar & Reinhart.
Pomfret, Richard (1988) *Unequal Trade.* Oxford: Basil Blackwell.
Postel, Sandra (1993) "The Politics of Water," *Worldwatch* 6, no. 4.
———— (1994) "Carrying Capacity: Earth's Bottom Line." In Lester R. Brown, ed. *State of the World.* New York: W.W. Norton.
Powelson, John P. (1977) "The Oil Price Increase: Impacts on Industrialized and Less-Developed Countries," *Journal of Energy and Development* (Autumn).
Rathjens, George (1995) "Rethinking Nuclear Proliferation," *Washington Quarterly* 18 (Winter).
Ravallion, Martin, and Shaohua Chen (1997) "What Can New Survey Data Tell Us About Recent Changes in Distribution and Poverty?" *World Bank Economic Review.* Washington, DC: World Bank.
Ray D. L., and L. Guzzo (1992) *Trashing the Planet.* New York: Harper Perennial.
Redfern, Paul (1995) "Africa: Left Out?" *East Africa* (October 30–November 5). Quoted in *World Press Review,* June 6, 1996.
Renan, Ernest (1996) "What Is a Nation?" In Geoff Eley and Ronald Grigor, eds. *Becoming National: A Reader.* New York: Oxford University Press.
Ricardo, David (1981) *Works and Correspondence of David Ricardo: Principles of Political Economy and Taxation.* London: Cambridge University Press.
Roberts, Brad (1995) *Weapons Proliferation in the 1990s.* Cambridge: MIT Press.
Rourke, John T. (1995) *International Politics on the World Stage.* Fifth edition. Guilford, CT: Dushkin Publishing Group.
———— (1997) *International Politics on the World Stage.* Sixth edition. Guilford, CT: Dushkin/McGraw Hill.
Russell, Dick (1995) "High-Seas Fishing: Lawless No Longer," *Amicus Journal* 17, no. 3.
Sagan, Scott D. (1986) "1914 Revisited: Allies, Offense, and Instability," *International Security* 11, no. 2.
Sarkar, Amin U., and Karen L. Ebbs (1992) "A Possible Solution to Tropical Troubles?" *Futures* 24, no. 7.
Scheffer, David J. (1996) "International Judicial Intervention," *Foreign Policy* 102 (Spring).

Schlesinger, James (1993) "The Impact of Nuclear Weapons on History," *Washington Quarterly* 16 (Autumn).

Schumacher, E. F. (1993) *Small Is Beautiful: Economics as if People Mattered*. San Francisco: Harper & Row.

Seis, M. (1996) "An Eco-critical Criminological Analysis of the 1990 Clean Air Act." Ph.D. diss., Indiana University of Pennsylvania.

Shilts, Randy (1987) *And the Band Played On*. New York: St. Martin's Press.

Simon, Julian L. (1990) *Population Matters: People, Resources, Environment, and Immigration*. New Brunswick, NJ: Transaction Publishers.

Sivard, Ruth Leger (1991) *World Military and Social Expenditures 1991*. Washington, DC: World Priorities.

Small, Melvin, and J. David Singer (1982) *Resort to Arms: International and Civil Wars, 1816–1980*. Beverly Hills, CA: Sage Publications.

Smith, Adam (1910) *An Inquiry into the Nature and Causes of the Wealth of Nations*. London: J.M. Dutton.

Steinberg, Gerald M. (1994) "U.S. Non-Proliferation Policy: Global Regimes and Regional Realities," *Contemporary Security Policy* 15, no. 3.

Sugar, Peter F., and Ivo John Lederer, eds. (1994) *Nationalism in Eastern Europe*. Seattle: University of Washington Press.

Switzer, Jacqueline Vaughn (1994) *Environmental Politics: Domestic and Global Dimensions*. New York: St Martin's Press.

Teich, Mikuláš, and Roy Porter, eds. (1993) *The National Question in Europe in Historical Context*. Cambridge: Cambridge University Press.

Toffler, Alvin, and Heidi Toffler (1991) "Economic Times Zones: Fast Versus Slow," *New Perspectives Quarterly* 8, no. 4.

Tumulty, Brian (1994) "U.S. Industry Confronts Cost of Implementing GATT," Gannett News Service, July 18.

Tyler, Patrick E. (1997) "China and Red Cross Agree to New Talks on Prison Visits," *New York Times*, February 29.

UN (United Nations) (1973) "Report of the Secretary-General on the Implementation of the Security Council Resolution 340." UN document S/11052/Rev., October 27.

—— (1976) "World Plan of Action, Report of the World Conference of the International Women's Year, Mexico City, June 19–July 1, 1975." F/Conf. 66/34. New York: United Nations.

—— (1988) *Human Rights: Questions and Answers*. New York: United Nations.

—— (1995) *Convention to Combat Desertification and Drought, Particularly in Africa*. New York: United Nations.

—— (1998) "United Nations Press Briefing on Kyoto Protocol," March 16.

UNACC/SCN (United Nations Administrative Committee on Coordination/Subcommittee on Nutrition (1997) "Update on the Nutrition Situation 1996: Summary of Results for the Third Report on the World Nutrition Situation." Geneva: ACC/SCN.

UNDESIPA (United Nations Department for Economic and Social Information and Policy Analysis, Population Division) (1994) "The Sex and Age Distribution of the World Populations." New York: United Nations.

—— (1995) "World Urbanization Prospects: The 1994 Revision." New York: United Nations.

UNDP (United Nations Development Programme) (1994) *Human Development Report*. New York: Oxford University Press.

—— (1996) *Human Development Report*. New York: Oxford University Press.

—— (1997) *Human Development Report*. New York: Oxford University Press.

UNDPI (United Nations Department of Public Information) (1990) *United Nations Peace-keeping.* UN: UNDPI: DPI/1048, May.
—— (1996) *The Blue Helmets: A Review of United Nations Peace-keeping.* Third edition. UN: UNDPI (DPI 1800, Sales No.: E.96.I.14).
UNEP (United Nations Environment Programme) (1996) *Global Biodiversity Assessment.* Nairobi: UNEP.
UNGA (United National General Assembly) (1947) Resolution 109 (II), October 21.
—— (1948) Resolution 186 (ES-1), May 14.
UNHCR (United Nations High Commissioner for Refugees) (1995) *The State of the World's Refugees 1995.* Oxford: Oxford University Press.
UNICEF (United Nations Children's Fund) (various years) *The State of the World's Children.* New York: Oxford University Press.
—— (1987) *The State of the World's Children 1987.* New York: Oxford University Press.
—— (1993a) *The Progress of Nations.* New York: UNICEF.
—— (1993b) *The State of the World's Children 1993.* New York: Oxford University Press.
—— (1996a) "Press Release, Secretary-General Reports Big Progress for Children." New York: UNICEF CF/DOC/PR/1996–24.
—— (1996b) *The Progress of Nations.* New York: UNICEF.
UNSC (United Nations Security Council) (1948) Resolution 801, May 29.
USBC (United States Bureau of the Census, International Programs Center) (1994) *World Population Profile.* Washington, DC: Government Printing Office.
—— (1997) "Money Income in the United States: 1996." *Current Population Reports,* pp. 60–197. Washington, DC: USBC.
USCC and AN (United States Code Congressional and Administrative News) (1991) 101st Congress, Second Session. Legislative History Clean Air Act Amendments, January, No. 10D. Minneapolis/St Paul: West Publishing.
USG (United States Government) (1995) *World Military Expenditures and Arms Transfers, 1993–94.* Washington, DC: Government Printing Office.
USGAO (United States General Accounting Office) (1991) "Child Labor: Characteristics of Working Children." Washington, DC: USGAO.
USSCEPW (United States Senate Committee on Environmental and Public Works) (1993) "Three Years Later: Report Card on the 1990 Clean Air Act Amendments, November 15." Washington, DC: Government Printing Office.
Valente, C. M., and W. D. Valente (1995) *Introduction to Environmental Law and Policy: Protecting the Environment Through Law.* Minneapolis/St. Paul: West Publishing.
Võ, X. H. (1994) *Oil, the Persian Gulf States, and the United States.* Westport, CT: Praeger.
von Laue, Theodore H. (1987) *The World Revolution of Westernization: The Twentieth Century in Global Perspective.* New York: Oxford University Press.
Warner, Sir Frederick (1991) "The Environmental Consequences of the Gulf War," *Environment* 33, no. 5.
WBCSD (World Business Council for Sustainable Development) (1996) *The Changing Future for Paper.* Geneva: WBCSD.
WCED (World Commission on Environment and Development) (1987) *Our Common Future.* Oxford: Oxford University Press.
Weber, Eugen (1964) *Varieties of Fascism: Doctrines of Revolution in the Twentieth Century.* Princeton, NJ: Van Nostrand.

Weber, P. (1993) "Reviving Coral Reefs." In Lester R. Brown, ed. *State of the World*. New York: W.W. Norton.

Weeks, John R. (1996) *Population: An Introduction to Concepts and Issues*. Belmont, CA: Wadsworth.

Weston, Burns H. (1992) "Human Rights." In Richard Pierre Claude and Burns H. Weston, eds. *Human Rights in the World Community*. Philadelphia: University of Pennsylvania Press.

WFA (World Federalist Association) (1996) *The Global Economy*. Part 2: *TNCs and Global Governance*. Washington, DC: WFA.

Wilkinson R. G. (1992) "Income Distribution and Life Expectancy," *British Medical Journal* 304, pp. 165–168.

Wiseberg, Laurie S. (1992) "Human Rights Nongovernmental Organizations." In Richard Pierre Claude and Burns H. Weston, eds. *Human Rights in the World Community*. Philadelphia: University of Pennsylvania Press.

Wiseman, Henry (1983) "United Nations Peacekeeping: An Historical Overview," In Henry Wiseman, ed. *Peacekeeping, Appraisals and Proposals*. New York: Pergamon Press.

World Bank (1993) *World Development Report*. Oxford: Oxford University Press.

――― (1994) *Investing in Infrastructure: World Development Report 1994*. New York: Oxford University Press.

――― (1995) *World Debt Tables*. Washington, DC: World Bank.

――― (1996) *From Plan to Market: World Development Report 1996*. New York: Oxford University Press.

――― (1997a) *The State in a Changing World: World Development Report 1997*. New York: Oxford University Press.

――― (1997b) *World Development Indicators*. New York: Oxford University Press.

WRI (World Resources Institute) (1992) *Global Biodiversity Strategy*. Washington, DC: World Resource Institute.

――― (1994) *World Resources 1994–95: A Guide to the Global Environment*. New York: Oxford University Press.

Zhu, B., et al. (1996) "Cigarette Smoking and Its Risk Factors Among Elementary School Students of Beijing," *American Journal of Public Health* 86, no. 3.

The Contributors

Elise Boulding is professor emerita of sociology at Dartmouth College and former secretary-general of the International Peace Research Association. She has undertaken numerous transnational and comparative cross-national studies on conflict and peace, development, and women in society. A scholar-activist, she was international chair of the Women's International League for Peace and Freedom in the late 1960s. Among her many publications are *Women in the Twentieth Century World,* 1977; *Building a Global Civic Culture: Education for an Interdependent World,* 1990; *The Underside of History: A View of Women Though Time,* 1992; *Building Peace in the Middle East: Challenges for States and Civil Society,* 1994; *The Future: Images and Processes,* with Kenneth Boulding, 1995.

Stephen Collett is director of the Quaker United Nations Office in New York and UN representative for the world body of Quakers, the Friends World Committee for Consultation. He has taught trade and development, geography, and international organizations at the College of Agder in Kristiansand, Norway, and at Earlham College in Richmond, Indiana. He is the author or coauthor of books and articles on international economics, sustainable development, and humanitarian assistance.

John K. Cox is assistant professor of history at Wheeling Jesuit University. His dissertation was a biography of the Slovene Communist theoretician Edvard Kardelj. His recent publications include an article on Josip Broz Tito and an interpretive essay on the rise of Marxist socialism. Cox has lectured to numerous academic and nonacademic audiences on the

279

wars in Bosnia and Croatia. He has traveled extensively in central Europe and the Balkans and is currently secretary-treasurer of the Society for Slovene Studies.

George Kent is professor of political science at the University of Hawaii at Manoa. He has written several books, including *The Political Economy of Hunger: The Silent Holocaust,* 1984; *Fish, Food, and Hunger: The Potential of Fisheries for Alleviating Malnutrition,* 1987; *The Politics of Children's Survival and Children in the International Political Economy,* 1991. He is coconvener of the Commission on International Human Rights of the International Peace Research Association and coordinator of the Global Task Force on Children's Nutrition Rights. He has worked as a consultant with the Food and Agriculture Organization of the United Nations, the United Nations Children's Fund, and several nongovernmental organizations.

Ellen Percy Kraly is a professor in the Department of Geography at Colgate University. She was a member of the National Academy of Sciences Panel on Immigration Statistics, 1983–1985, and has conducted research for the United Nations Statistical Commission, the U.S. Immigration and Naturalization Service, and the U.S. Commission on Immigration Reform. She has been president of the Population Specialty Group of the Association of American Geographers and is author of numerous articles on immigration and U.S. population growth and other migration issues.

Jeffrey S. Lantis is assistant professor of political science at the College of Wooster. He has taught a variety of international relations courses and served as associate editor for *Brassey's Mershon American Defense Annual 1995/6.* His recent publications have appeared in *German Politics and Society* and *International Studies Notes.* He is also the recent author of *Domestic Constraints and the Breakdown of International Agreements.*

Bruce E. Moon is a professor in the Department of International Relations at Lehigh University. He is author of *The Political Economy of Basic Human Needs,* 1991, and *Dilemmas of International Trade,* 1996. His articles have appeared in *International Studies Quarterly, International Organization,* the *American Journal of Political Science, Comparative Political Studies, Political Research Quarterly,* and the *Journal of Conflict Resolution.* His research in international political economy and foreign policy has also appeared in several edited volumes.

Marjorie E. Nelson is a faculty member of the Department of Family Medicine at the Ohio University College of Osteopathic Medicine and is currently head of the Preventive Medicine and Public Health Section there. After residency training in Philadelphia, she was staff physician with the American Friends Service Committee Rehabilitation Project at Quang Ngai Hospital in Vietnam from 1967 to 1969. She has been a local health officer and medical director of a Planned Parenthood affiliate, and has worked with the Hospital Ship HOPE in Guinea, West Africa. She has traveled widely in Asia, Africa, and Europe.

Don Reeves has recently retired as economic policy analyst at Bread for the World (BFW) Institute, which engages Christians and other concerned citizens in educational activities on hunger- and poverty-related issues. On the BFW Institute staff from 1987–1998, he directed a church- and farm-sponsored educational effort on issues of U.S. agriculture, trade, and development during the early 1990s. From 1977 to 1980, he was legislative secretary for the Friends Committee on National Legislation, the Quaker lobby group in Washington, D.C. He is a founding chair of the Nebraska Farm Crisis Response Program under Interchurch Ministries of Nebraska.

Gerald W. Sazama is associate professor of economics at the University of Connecticut. He was a Fulbright fellow for research in energy economics in Costa Rica and held a Social Science Research Council fellowship to research land taxation in Chile. He was also a consultant on project evaluation and taxation for USAID in Bolivia, Costa Rica, and Nicaragua, and has done consultation work with the World Bank on a training project in Afghanistan. He has published in the *National Tax Journal, Journal of Regional Science, Europe-Asia Studies,* and *Economic Development and Cultural Change.*

Karrin Scapple is assistant professor of political science at Southwest Missouri State University. Her areas of specialty include international environmental politics and policy, international law, and international organizations. Her current research interests include defining effectiveness in environmental treaties, Antarctica, and sustainable development. She has published in *International Studies Notes* and the *American Society of International Law Antarctic Interest Group Newsletter.*

Mark Seis is assistant professor of sociology at Fort Lewis College in Durango, Colorado. His dissertation was an ecological policy critique of the 1990 Clean Air Act (CAA). His primary research interests are environmental crime and law and policy. He has published on various topics of the environment ranging from the CAA to a Native American perspective of environmental crime. He is coauthor of a textbook, *A Primer in Environmental Crime.*

D. Neil Snarr is professor of sociology and director of international education at Wilmington College (Ohio). He also directs the Global Issues program, a three-hour course required of all freshmen and seniors. He has traveled extensively, taken student groups abroad, initiated international programs for students, published in major journals, and edited several books.

Michael T. Snarr is assistant professor of political science and codirector of international studies at Wheeling Jesuit University. His research focuses on Latin American foreign policy toward the United States. His travels to the developing world have included trips with students to central Mexico as part of a course on contemporary Mexico.

Carolyn M. Stephenson is associate professor of political science at the University of Hawaii at Manoa. She is editor of the book *Alternative Methods for International Security,* 1982, and author of the forthcoming *Common Sense and the Common Defense.* She was director of peace studies at Colgate University, and was coeditor of *Peace and Change: A Journal of Peace Research* for a number of years. The author of many articles on the development of peace studies, she is currently doing research on nongovernmental women's, environmental, and disarmament organizations at United Nations conferences.

Index

About the Book

Introducing students to today's most pressing global issues, this text explores the various dimensions of conflict and security, the global economy, development, and the environment.

The material is designed to be easily accessible to readers with little or no prior knowledge of the topics covered. Each chapter provides an analytical overview of the issue addressed, identifies the central actors and perspectives, and outlines progress made in the area (as well as prospects for the future). The book is enriched by challenging questions posed to enhance students' appreciation of the complexities involved, as well as by suggestions for further reading.

Michael T. Snarr is assistant professor of political science and codirector of international studies at Wheeling Jesuit University. **D. Neil Snarr** is professor of sociology and director of the Global Issues program at Wilmington College.